Electron Diffraction and Structural Imaging

Electron Diffraction and Structural Imaging

Guest Editors

Partha Pratim Das
Arturo Ponce-Pedraza
Enrico Mugnaioli
Stavros Nicolopoulos

MDPI

Basel • Beijing • Wuhan • Barcelona • Belgrade • Novi Sad • Cluj • Manchester

Guest Editors
Partha Pratim Das
NanoMEGAS SPRL
Brussels
Belgium

Arturo Ponce-Pedraza
Department of Physics & Astronomy
University of Texas at San Antonio
San Antonio, TX
USA

Enrico Mugnaioli
Department of Earth Sciences
University of Pisa
Pisa
Italy

Stavros Nicolopoulos
NanoMEGAS SPRL
Brussels
Belgium

Editorial Office
MDPI AG
Grosspeteranlage 5
4052 Basel, Switzerland

This is a reprint of the Special Issue, published open access by the journal *Symmetry* (ISSN 2073-8994), freely accessible at: https://www.mdpi.com/journal/symmetry/special_issues/Electron_Diffraction_Structural_Imaging.

For citation purposes, cite each article independently as indicated on the article page online and as indicated below:

Lastname, A.A.; Lastname, B.B. Article Title. *Journal Name* **Year**, *Volume Number*, Page Range.

ISBN 978-3-7258-5371-7 (Hbk)
ISBN 978-3-7258-5372-4 (PDF)
https://doi.org/10.3390/books978-3-7258-5372-4

Contents

About the Editors

Partha Pratim Das

Partha Pratim Das is an Application Scientist at NanoMEGAS SPRL, Belgium. He earned his M.S. in Chemical Sciences from the Indian Institute of Science, Bangalore, and his Ph.D. in Materials Science from the University of Zurich in 2012, where he focused on structural disorder analysis using diffuse scattering from single-crystal X-ray data. Since then, at NanoMEGAS, he has contributed to the development of precession-assisted 3D electron diffraction for solving complex nanostructures, electron pair distribution function mapping for amorphous materials, and novel methodologies for structural analysis. His research focuses on advanced electron diffraction approaches for beam-sensitive materials, in situ experiments, cultural heritage studies, and the investigation of amorphous and nanocrystalline systems. He has authored more than 50 publications with nearly 700 citations, co-organized over 30 international workshops including IUCr2017 in Hyderabad, and recently co-supervised a Ph.D. student from ITQ Valencia, Spain.

Arturo Ponce-Pedraza

Arturo Ponce Pedraza is Professor and Department Chair of Physics and Astronomy at the University of Texas at San Antonio, where he leads the Structure Physics and Electron Microscopy group. He earned his MS in Solid State Physics from the University of Puebla, Mexico, and his PhD in Materials Science and Engineering from the University of Cadiz, Spain, in 2003. He carried out postdoctoral research at UNAM, Mexico, and at CNRS, France, between 2005 and 2007, focusing on electron crystallography of interfaces. From 2008 to 2011, he was Senior Scientist and head of the microscopy center at CIQA, Saltillo, Mexico. He joined UTSA in 2011 as Director of the Kleberg Advanced Microscopy Center and has been faculty since 2013. He is the author of more than 120 peer-reviewed articles, in journals including *Science*, *Nature*, and *ACS Nano*, and has supervised multiple doctoral and postdoctoral researchers. He is a member of MSA, MRS, and AMMM, and was inducted into the Mexican Academy of Science in 2018. His research spans electron diffraction, aberration-corrected TEM, holography, and in situ electron microscopy.

Enrico Mugnaioli

Enrico Mugnaioli is an Associate Professor of Mineralogy at the University of Pisa, Italy, where he manages the HR-TEM CISUP laboratory and provides TEM expertise for internal and external collaborations. He earned his bachelor's and master's degrees cum laude and his PhD in Earth Sciences from the University of Siena, where he applied electron crystallography methods for the characterization of serpentine minerals. From 2007 to 2014, he worked in Prof. U. Kolb's group at the University of Mainz, Germany, and was one of the first developer of three-dimensional electron diffraction (3DED). Between 2016 and 2022, he was Senior Scientific Collaborator at the Italian Institute of Technology@NEST in Pisa, developing low-dose diffraction protocols for beam-sensitive materials. He is currently focusing on the application of 3DED to different microcrystalline materials, among which hybrid compounds, high-tech devices, nanoparticles, rare minerals, extraterrestrial samples, environmental specimens, porous materials and pharmaceuticals. He has authored more than 140 peer-reviewed papers with over 4400 citations.

Stavros Nicolopoulos

Stavros Nicolopoulos obtained his PhD in Materials Science in 1989 from the University of Grenoble, France, with a thesis on X-ray and neutron crystallography of magnetic structures. From 1991 to 1996, he was Associate Professor at the Complutense University of Madrid, Spain, focusing on biomaterials research. Between 1996 and 2004, he worked at Philips Electron Optics as a TEM applications specialist and later served as President of FEI in Spain. Since 2004, he has been Director and Co-Founder of NanoMEGAS SPRL, Brussels, a global leader in 4D-STEM precession electron diffraction instrumentation. His team developed the Digistar precession device and the ASTAR tool for phase and orientation imaging in TEM. He has authored more than 95 peer-reviewed publications with over 3300 citations and is co-inventor on five international patents. In 2004, he was appointed Consultant to the IUCr Commission on Electron Crystallography, and in 2011 he received the *Microscopy Today* Innovation Award for the ASTAR system, recognized as one of the ten most important inventions in electron microscopy that year. His research interests include advanced electron diffraction/crystallography, in situ TEM, and archaeometry.

Preface

Electron diffraction (ED) and structural imaging have re-emerged as vital tools for structural science, offering atomic-scale insights across physics, chemistry, materials science, and biology. This Reprint brings together contributions that demonstrate the breadth of ED applications, from energy storage materials and metallic alloys to steels, intermetallics, oxides, thin films, and framework compounds. The scope extends from structure analysis to methodological studies that address challenges beyond the reach of conventional X-ray techniques.

Our aim in assembling this collection was to showcase both the maturity of established approaches and the promise of emerging ones, highlighting how ED continues to evolve into a versatile and indispensable technique. This Reprint is addressed to scientists and students alike, providing both a reference for experienced researchers and a gateway for newcomers entering this rapidly growing field.

Partha Pratim Das, Arturo Ponce-Pedraza, Enrico Mugnaioli, and Stavros Nicolopoulos
Guest Editors

symmetry

MDPI

Editorial

Special Issue: Electron Diffraction and Structural Imaging—Volume I

Partha Pratim Das [1,*], Arturo Ponce-Pedraza [2], Enrico Mugnaioli [3] and Stavros Nicolopoulos [1]

1 NanoMEGAS SPRL, Rue Èmile Claus 49 bte 9, 1050 Brussels, Belgium; info@nanomegas.com
2 Department of Physics and Astronomy, The University of Texas at San Antonio, San Antonio, TX 78249, USA; arturo.ponce@utsa.edu
3 Department of Earth Sciences, University of Pisa, Via S. Maria 53, 56126 Pisa, Italy; enrico.mugnaioli@unipi.it
* Correspondence: partha@nanomegas.com

In recent years, electron diffraction (ED) and structural imaging have undergone a major resurgence in the scientific community, driven by continuous advancements in transmission electron microscopy (TEM) instrumentation, such as Cs correctors, direct detection cameras and automation, and the development or expansion of analytical methods, such as cryo-EM, beam precession, 4D Scanning Electron Diffraction, 3D electron diffraction, 4D-STEM, and ptychography. These innovations have enabled the atomic-scale structural characterization of a wide range of nanocrystalline materials, spanning functional materials, zeolites, energy storage materials, minerals, organics, pharmaceuticals, and proteins.

One of the key strengths of ED lies in its low-dose capabilities compared to conventional imaging techniques, allowing for the study of beam-sensitive and nanocrystalline materials that would otherwise be inaccessible by laboratory or synchrotron X-ray techniques. Today, ED is widely applied in atomic structure determination, phase/orientation/strain mapping, electric field analysis, and even in the study of amorphous materials. Furthermore, the integration of in situ TEM holders (gas, liquid, and heating holders) has allowed the analysis of materials in real-time and near-natural conditions. All these applications benefit from the understanding and exploitation of symmetry at the atomic scale, from conventional space groups to higher-dimensional formalisms in incommensurate structures and dynamic symmetry evolution under external stimuli [1–13].

This Special Issue was published in two parts. Volume I includes 10 articles authored by international researchers, while Volume II features 9 original contributions that highlight the versatility, depth, and ongoing innovation in the field of electron diffraction and structural imaging.

The first paper by Orlova et al. presents a comprehensive study of LiNi0.8Mn0.1Co0.1O2 (NMC811) cathode materials using a combination of PXRD, HAADF-STEM, and electron diffraction tomography (EDT). The authors investigate cation disorder, particularly Li+/Ni2+ mixing, which strongly influences electrochemical performance. The defect formation mechanisms were studied through Rietveld refinement and various structure-sensitive metrics, while local variations were mapped using STEM. This multi-scale approach provides a full picture of disorder and demonstrates how complementary techniques can bridge the gap between bulk structure and local defect environments [14].

The second paper by Rauch et al. reviews the development and capabilities of precession-assisted ACOM-TEM (commercially known as ASTAR)—an automated tool for electron diffraction pattern indexing and orientation mapping in TEM using scanning precession electron diffraction data. Initially designed as an alternative to EBSD and XRD pole figures with nanometer-scale spatial resolution, ACOM/TEM has since evolved into a powerful platform for phase identification and crystallographic analysis in TEM and is used in many material science labs. The paper addresses key technical challenges such as indexing complex diffraction patterns with overlapping contributions from multiple phases and resolving 180° orientation ambiguities. Future perspectives are outlined, including the 3D reconstruction of crystallographic phases within materials, further expanding applications in the field of advanced materials [15].

The third paper by Quintelier et al. applies precession electron diffraction (PED) and 3DED to the study of Li1.2Ni0.13Mn0.54Co0.13O2 cathodes used in lithium-ion batteries. Due to the nanoscale structured domains and the beam sensitivity, traditional XRD and imaging fall short in quantifying phase changes. The authors developed a method to determine the volume ratio of spinel-like phases of honeycomb structures using low-dose electron diffraction. After 150 charge/discharge cycles, 4% of the volume converted irreversibly, providing insight into the degradation mechanisms in high-capacity cathodes [16].

The fourth paper by Yang et al. investigates whether cryogenic conditions are essential for 3DED of small organic molecules. The structures of sucrose and a newly reported compound, azobenzene tetracarboxylic acid (H4ABTC), were determined at both room and cryogenic temperatures. Despite common assumptions, the results show that when samples are stable in vacuum, room temperature data can yield structures of comparable quality, including hydrogen atom positions. The study emphasizes merging data as a key strategy to improve resolution and reliability [17].

The fifth paper by Tamari et al. resolves a long-standing debate regarding symmetry in high-temperature "Al3Mn" T-phase alloys extended to the Al–Mn–Pt system. Using crystallographic and metallurgical evidence, the authors determine that the Al-rich phase is non-centrosymmetric (Pna21) while the Al-poor variant is centrosymmetric (Pnam). This work highlights the correlation between composition and symmetry in complex metallic alloys and provides a framework for further investigations [18].

The sixth paper by Mythili et al. focuses on retained austenite in reduced activation ferritic martensitic (RAFM) steels used in nuclear fusion reactors. The combination of synchrotron XRD, Mössbauer spectroscopy, and precession-assisted ACOM-TEM/ASTAR reveals a subtle retained austenite that is not visible with conventional methods. This study underscores the importance of microstructural homogeneity in optimizing the mechanical properties of RAFM steels [19].

The seventh paper by Klein et al. introduces precession-assisted low-dose electron diffraction tomography (LD-EDT) as a structure solution method for beam-sensitive nanocrystals. Examples include two synthetic oxides, a mineral, and an MOF, all solved and refined using dynamical diffraction data at doses below 0.1 $e^-/\text{Å}^2$. This method enables the reliable structure determination of complex materials previously inaccessible by X-rays or traditional ED approaches [20].

The eighth article by Singh et al. examines the creep behavior of Nb- and Ta-rich γ-TiAl alloys using a suite of electron diffraction and imaging techniques, including precession-assisted ACOM-TEM/ASTAR, EBSD, and TEM. Creep tests at high temperatures reveal distinct microstructural instabilities and intermetallic phase formation (e.g., τ phase) in Ta-rich alloys. The results highlight how alloying elements influence phase evolution under stress [21].

The ninth article by Yang et al. explores Ruddlesden–Popper (RP) fault structures and interface chemistry in NdNiO3 thin films grown via molecular beam epitaxy. Using aberration-corrected STEM and spectroscopy, the authors show that elemental intermixing and local strain lead to Ni valence variations at the faults. This microstructural understanding is essential for tuning functional properties in nickelates [22].

The tenth paper of the first volume by Das et al. provides a systematic evaluation of global optimization methods for solving structures from incomplete 3DED datasets collected with and without precession. In cases where traditional methods fail due to limited tilt range, beam damage, or dynamic scattering, the proposed approach enables successful structure determination. This study benchmarks global optimization against conventional structure solution paths, showing that it is a reliable alternative for difficult cases [23].

Conflicts of Interest: The authors declare no conflicts of interest.

References

1. Haider, M.; Uhlemann, S.; Schwan, E.; Rose, H.; Kabius, B.; Urban, K. Electron microscopy image enhanced. *Nature* **1998**, *392*, 768–769. [CrossRef]
2. McMullan, G.; Faruqi, A.R.; Henderson, R. Direct electron detectors. *Methods Enzymol.* **2016**, *579*, 1–17.
3. Mastronarde, D.N. Automated electron microscope tomography using robust prediction of specimen movements. *J. Struct. Biol.* **2005**, *152*, 36–51. [CrossRef]
4. Nogales, E. The development of cryo-EM into a mainstream structural biology technique. *Nat. Methods* **2016**, *13*, 24–27. [CrossRef] [PubMed]
5. Vincent, R.; Midgley, P.A. Double conical beam-rocking system for measurement of integrated electron diffraction intensities. *Ultramicroscopy* **1994**, *53*, 271–282. [CrossRef]
6. Ophus, C. Four-Dimensional Scanning Transmission Electron Microscopy (4D-STEM): From Scanning Nanodiffraction to Ptychography and Beyond. *Microsc. Microanal.* **2019**, *25*, 563–582. [CrossRef]
7. Chen, Z.; Jiang, Y.; Shao, Y.-T.; Holtz, M.E.; Odstrčil, M.; Guizar-Sicairos, M.; Han, Y.; Gao, P.; Tate, M.W.; Gruner, S.M.; et al. Electron ptychography achieves atomic-resolution limits set by lattice vibrations. *Science* **2021**, *372*, 826–831. [CrossRef]
8. Gemmi, M.; Mugnaioli, E.; Gorelik, T.E.; Kolb, U.; Palatinus, L.; Boullay, P.; Hovmöller, S.; Abrahams, J.P. 3D Electron Diffraction: The Nanocrystallography Revolution. *ACS Cent. Sci.* **2019**, *5*, 1315–1329. [CrossRef]
9. Taheri, M.L.; Stach, E.A.; Arslan, I.; Crozier, P.A.; Kabius, B.C.; LaGrange, T.; Minor, A.M.; Takeda, S.; Tanase, M.; Wagner, J.B.; et al. Current status and future directions for in situ transmission electron microscopy. *Ultramicroscopy* **2016**, *170*, 86–95. [CrossRef]
10. Viladot, D.; Véron, M.; Gemmi, M.; Nicolopoulos, S.; Buffat, P.; Lefebvre, W.; Urban, K.; Midgley, P.A. Orientation and phase mapping in the transmission electron microscope using precession-assisted diffraction spot recognition: State-of-the-art results. *J. Microsc.* **2013**, *252*, 23–24. [CrossRef]
11. Choi, S.-H.; Sim, J.-M.; Shin, J.; Ryu, S.-H.; Hwang, T.; Lim, S.Y.; Oh, H.-J.; Kwag, J.-H.; Lee, J.-Y.; Song, K.-C.; et al. Unveiling the Hybrid-Channel (poly-Si/IGO) Structure for 3D NAND Flash Memory for Improving the Cell Current and GIDL-Assisted Erase Operation. *Small Struct.* **2025**, *6*, 2400495. [CrossRef]
12. Tirado, J.I.; Das, P.P.; Jorda, J.L.; Palatinus, L.; Plana-Ruiz, S.; Simancas, J.; Simancas, R.; Nicolopoulos, S.; Valencia, S.; Rey, F. Structure Determination of As-Made Zeolite ITQ-52 by Three-Dimensional Electron Diffraction. *Microporous Mesoporous Mater.* **2025**, *382*, 113392. [CrossRef]
13. Rakita, Y.; Hart, J.L.; Das, P.P.; Shahrezaei, S.; Foley, D.L.; Mathaudhu, S.N.; Nicolopoulos, S.; Taheri, M.L.; Billinge, S.J.L. Mapping Structural Heterogeneity at the Nanoscale with Scanning Nano-Structure Electron Microscopy (SNEM). *Acta Mater.* **2023**, *242*, 118426. [CrossRef]
14. Orlova, E.D.; Savina, A.A.; Abakumov, S.A.; Morozov, A.V.; Abakumov, A.M. Comprehensive Study of Li^+/Ni^{2+} Disorder in Ni-Rich NMCs Cathodes for Li-Ion Batteries. *Symmetry* **2021**, *13*, 1628. [CrossRef]
15. Rauch, E.F.; Harrison, P.; Zhou, X.; Herbig, M.; Ludwig, W.; Véron, M. New Features in Crystal Orientation and Phase Mapping for Transmission Electron Microscopy. *Symmetry* **2021**, *13*, 1675. [CrossRef]
16. Quintelier, M.; Perkisas, T.; Poppe, R.; Batuk, M.; Hendrickx, M.; Hadermann, J. Determination of Spinel Content in Cycled $Li_{1.2}Ni_{0.13}Mn_{0.54}Co_{0.13}O_2$ Using Three Dimensional Electron Diffraction and Precession Electron Diffraction. *Symmetry* **2021**, *13*, 1989. [CrossRef]
17. Yang, T.; Waitschat, S.; Inge, A.K.; Stock, N.; Zou, X.; Xu, H. A Comparison of Structure Determination of Small Organic Molecules by 3D Electron Diffraction at Cryogenic and Room Temperature. *Symmetry* **2021**, *13*, 2131. [CrossRef]
18. Tamari, R.; Grushko, B.; Meshi, L. Electron Diffraction Study of the Space Group Variation in the Al–Mn–Pt T-Phase. *Symmetry* **2022**, *14*, 38. [CrossRef]
19. Mythili, R.; Kirana, R.; Singh, L.H.; Govindaraj, R.; Sinha, A.K.; Singh, M.N.; Saroja, S.; Vijayalakshmi, M.; Deb, S.K. Identification of retained austenite in 9Cr-1.4W-0.06Ta-0.12C reduced activation ferritic martensitic steel. *Symmetry* **2022**, *14*, 196. [CrossRef]

20. Klein, H.; Kodjikian, S.; Yörük, E.; Bordet, P. Low-Dose Electron Crystallography: Structure Solution and Refinement. *Symmetry* **2022**, *14*, 245. [CrossRef]
21. Singh, V.; Mondal, C.; Sarkar, R.; Roy, S.; Omprakash, C.M.; Ghosal, P. Characterization of Microstructure of Crept Nb and Ta-Rich γ-TiAl Alloys by Automated Crystal Orientation Mapping and Electron Back Scatter Diffraction. *Symmetry* **2022**, *14*, 399. [CrossRef]
22. Yang, C.; Wang, Y.; Putzky, D.; Sigle, W.; Wang, H.; Ortiz, R.A.; Logvenov, G.; Benckiser, E.; Keimer, B.; van Aken, P.A. Ruddlesden–Popper faults in $NdNiO_3$ thin films. *Symmetry* **2022**, *14*, 464. [CrossRef]
23. Das, P.P.; Plana-Ruiz, S.; Galanis, A.S.; Stewart, A.; Karavasili, F.; Nicolopoulos, S.; Putz, H.; Margiolaki, I.; Calamiotou, M.; Iezzi, G. Structure determination feasibility of three-dimensional electron diffraction in case of limited data. *Symmetry* **2022**, *14*, 2355. [CrossRef]

symmetry MDPI

Article

New Features in Crystal Orientation and Phase Mapping for Transmission Electron Microscopy

Edgar F. Rauch [1,*], Patrick Harrison [1], Xuyang Zhou [2], Michael Herbig [2], Wolfgang Ludwig [3] and Muriel Véron [1]

1 SIMAP, Grenoble INP, Université Grenoble Alpes, CNRS, 38000 Grenoble, France; patrick.harrison@simap.grenoble-inp.fr (P.H.); muriel.veron@grenoble-inp.fr (M.V.)
2 Max-Planck-Institut für Eisenforschung GmbH, Max-Planck-Str. 1, 40237 Düsseldorf, Germany; x.zhou@mpie.de (X.Z.); m.herbig@mpie.de (M.H.)
3 MATEIS, INSA Lyon, Université Lyon I, CNRS UMR 5510, 69621 Villeurbanne, France; ludwig@esrf.fr
* Correspondence: edgar.rauch@grenoble-inp.fr

Abstract: ACOM/TEM is an automated electron diffraction pattern indexing tool that enables the structure, phase and crystallographic orientation of materials to be routinely determined. The software package, which is part of ACOM/TEM, has substantially evolved over the last fifteen years and has pioneered numerous additional functions with the constant objective of improving its capabilities to make the tremendous amount of information contained in the diffraction patterns easily available to the user. Initially devoted to the analysis of local crystallographic texture, and as an alternative to both X-ray pole figure measurement and EBSD accessories for scanning electron microscopes, it has rapidly proven itself effective to distinguish multiple different phases contained within a given sample, including amorphous phases. Different strategies were developed to bypass the inherent limitations of transmission electron diffraction patterns, such as 180° ambiguities or the complexity of patterns produced from overlapping grains. Post processing algorithms have also been developed to improve the angular resolution and to increase the computing rate. The present paper aims to review some of these facilities. On-going works on 3D reconstruction are also introduced.

Keywords: ACOM/TEM; ASTAR; transmission electron microscopy; scanning precession electron diffraction (SPED); phase mapping; electron crystallography

1. Introduction

Automated crystal orientation mapping tools (ACOM) are largely used in electron microscopy to quantitatively analyze grain size and crystallographic textures of materials at micron or sub-micron scales. Among them, electron backscatter diffraction (EBSD) attachments for scanning electron microscopes (SEM) are the most popular and benefit from long-standing improvements, today reaching impressive speeds and indexing qualities. Similar developments for transmission electron microscopes (TEM) are more recent and are continuously gaining interest from the community. As their ACOM/SEM counterparts, the ACOM/TEM tools scan a small probe over the area of interest and acquire the complete set of local diffraction patterns producing a 4D dataset. For TEM operators, the characterization procedure is deeply modified as the data is stored on a disk so that the majority of the analysis is performed off-line, by extracting the relevant signal out of the stack of 'images' collected on-line. In addition to the calculation of orientation or phase mapping, this offers the possibility to highlight structural features (grains, twins, precipitates, dislocations, boundaries, etc.) with additional facilities, such as the construction of virtual images (bright- and dark-field, or computer processed combined images) and correlation coefficient maps. Off-line analysis also permits the indexing parameters to be optimized or the data to be corrected by using cleaning tools as the ones available in EBSD software, with no extra line cost.

The ACOM/TEM approach considered here has given rise to a software plus hardware package, known as the ASTAR^{TM} tool, which has been described on different occasions [1,2]. Its main features are the use of precession electron diffraction (PED) to improve the quality of the acquired patterns and the pioneering introduction of the template matching strategy for rapid orientation and phase identification. It is composed of four dedicated numerical software programs devoted, respectively, to pattern acquisition, template generation, pattern indexing, and results analysis. This package has significantly evolved over fifteen years by introducing novel functions and user-friendly facilities that increase the possibilities to analyze challenging materials and extract faint information from the diffracting signal. Electron precession is generated with the Digistar equipment supplied by NanoMEGAS SPRL [3]. ASTAR was successfully applied to a wide range of systems, including energy storage materials [4,5], amorphous materials [6,7], and nanomaterials [8,9].

The main objective of the present paper is to review the most salient improvements and to highlight the capabilities, as well as the limitations, of the strategies incorporated, or to be incorporated, into ASTAR. The template matching strategy is detailed and commented on in Section 2. In Section 3, it is demonstrated that this strategy remains efficient even when the diffraction patterns are acquired with a low-resolution camera. Section 4 describes how the 180° ambiguity problem, inherently associated to transmission election diffraction, is corrected. The final section introduces the ongoing works on 3D reconstructions that requires the diffracting signatures of all the overlapping grains to be distinguished.

2. Template Matching: Practical Aspects

The diffraction patterns collected in the TEM under scanning mode on polycrystalline materials are typically very far from zone axis (ZA) and correspond to random cuts of the reciprocal lattice by the Ewald sphere. In this case, the classical procedure that consisted of recognizing ZA patterns by manual comparison with a limited number of charts can no longer be exploited. However, it can be extended numerically to any orientation by pre-computing all possible patterns for all orientations of all the phases expected in the sample. Each of these theoretical patterns, called templates, is cross correlated to each of the collected patterns and the template giving the best score is selected, providing the desired information at a given location on the sample.

The cross-correlation strategy, known as template matching, uses a correlation index which is calculated as:

$$Q(i) = \frac{\sum_{j=1}^{m'} P(x_j, y_j) T_i(x_j, y_j)}{\sqrt{\sum_{j=1}^{m} P^2(x_j, y_j)} \sqrt{\sum_{j=1}^{m'} T_i^2(x_j, y_j)}} \tag{1}$$

In this equation, the diffraction pattern is represented by the intensity function $P(x,y)$ and every template i is given by the function $T_i(x,y)$. The pattern and template contain m pixels but of interest is the fact that the template only has a limited number of non-zero pixels m'.

2.1. Algorithm Improvements

Indeed, *template* matching implies that only a small fraction of the pattern will match the template. In the current approach, the theoretical pattern does not reproduce the entire diffracting signal but only the most relevant part of it. More precisely, a template contains typically less than $m' \sim 100$ datapoints $T_i(x,y)$, each being representative of the position and the intensity of one reflection. This differs from a full cross-correlation approach where the entire simulated pattern is reproduced and compared to the raw data [10]. The processing time is tremendously reduced by considering such a minimal representation of the pattern. For example, even if the size of the collected pattern is not more than 144 × 144 pixels—the standard size used in the present approach—the number of operations in Equation (1) is reduced from $m \sim 2.10^4$ (total number of pixels) down to $m' \sim 100$ (non-zero pixels in the template).

Additionally, the best match is related to the highest Q value for a given pattern. This means that the following equation would lead to exactly the same hierarchy and the same solution as Equation (1):

$$Q_{eff}(i) = \frac{\sum_{j=1}^{m'} P(x_j, y_j) T_i(x_j, y_j)}{\sqrt{\sum_{j=1}^{m'} T_i^2(x_j, y_j)}} \tag{2}$$

Because the selection of the best template for a given diffraction pattern does not depend on the normalization of the pattern, this calculation is omitted to speed up the template matching process. The drawback of such a simplification is that the correlation index (Q_{eff}) has an arbitrary value which is no longer bounded to a maximum score of 1. This results in an index score that is sensitive to the pattern brightness, but this has no impact on the results as each pattern is treated independently.

Templates are computed in a straightforward manner: the positions of the reflections are determined by a simple geometrical construction that depends on the lattice parameters and the Euler angles, whilst the template intensities are proportional to the square of the structure factor:

$$F_{hkl} = \sum_i f_i e^{2\pi i (hx_i + ky_i + lz_i)} \tag{3}$$

This requires the knowledge of the type and position of all atoms in the unit cell. The atomic scattering factor f_i is approximated by the atomic number. Moreover, the intensity dependence on the Bragg angle is merely omitted. This choice reinforces the weight of the reflections appearing at the rim of the pattern, the ones that are more sensitive to the exact orientation of the crystal, therefore, increasing the angular resolution.

2.2. Background Subtraction

If no energy filter is used, the raw data displays a significant background intensity due to inelastically scattered electrons that increases when approaching the transmitted beam. Such background is weakly orientation or phase dependent but will produce a substantial contribution to the Q value for all templates whatever the location of the theoretical reflections. Therefore, it is necessary to eliminate it. This is performed by subtracting the local background intensity at every pixel in the pattern. In practice, this is calculated by averaging the intensity over a circle of a given radius drawn around the current position. The radius is ideally half the smallest distance existing between the reflections. Negative values produced by this calculation are subsequently set to zero. Consequently, the transformed pattern contains relative intensities that are far more appropriate for template matching (Figure 1).

Additional image processing steps may be used to further optimize the phase and/or orientation recognition. Experience has shown that a threshold of around 5% of the maximum intensity as well as a gamma correction of 0.5—enhancing intermediate intensities—usually improve the results.

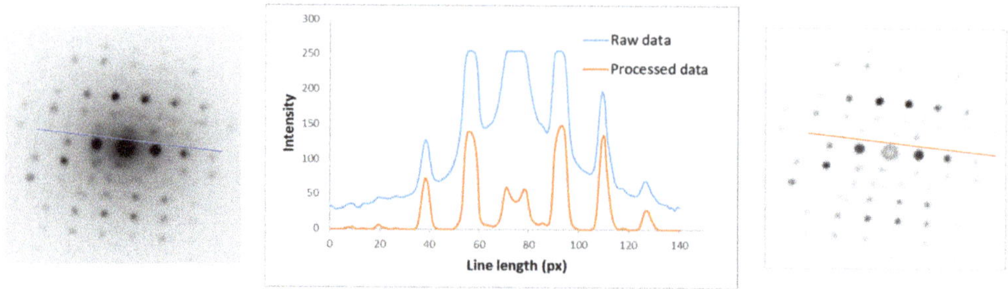

(a) (b) (c)

Figure 1. Background subtraction: (**a**) the raw diffraction pattern contains 144 × 144 pixels, (**b**) the intensity profiles before and after image processing are compared, (**c**) relative intensities are obtained by subtracting the background intensity computed around the current location with a radius of 5 pixels; Note that the transmitted beam is damaged because its size exceeds the selected radius.

3. Raw Pattern Size

As mentioned above, templates are crude estimates of the real diffraction patterns. They do not consider refined features like dynamical effects of the reflection profile and contain a very limited number of points. They may be considered as the diffraction pattern equivalent as minutiae for fingerprint representation. A precession angle of not more than 1° has proven very efficient to clean the pattern from strongly orientation-sensitive fluctuations (dynamical effects but also Kikuchi lines) [11]. Of interest is the fact that the orientation resolution is not bounded by the precession angle as long as the templates are corrected for precession [11–13].

In the standard procedure, the raw patterns are acquired on-line with an optical camera placed outside the TEM column. When scanning at high scan speeds, which may be hundreds of frames per second, the collected intensities are low such that binning (i.e., adding intensities of neighbouring pixels) is systematically performed. This substantially reduces the pattern resolution. The current trend in electron microscopy is to introduce new detectors which are faster and more sensitive but that also contain more pixels, up to several millions. While there is no doubt that orientation and phase mapping will benefit from these recent evolutions in hardware, this also comes with a tremendous increase in data size and acquisition time. To this respect, acquiring high resolution diffraction patterns is questionable, at least when templates are used to recognize the pattern. Figure 2 compares orientation and phase maps generated with patterns acquired with a 4k × 4k CMOS detector (TemCam-XF416-TVIPS) binned on-line to 512 × 512 pixels and further binned numerically to 256 × 256, 128 × 128, 64 × 64 and even 32 × 32 pixels for the sake of comparison.

The sample is extracted from a hypereutectoid pearlitic wire cold drawn to a true strain of ε = 6.02 and annealed for 2 min at 400 °C. Consequently, the ferritic grains contain substantial misorientations that are easily detected and even, surprisingly, for a reduction of the pattern resolution to just 32 × 32 pixels. At the lowest pattern resolutions, however, the ability to accurately identify intergranular cementite particles was lost. This is, however, an expected result as the distance between the closest reflections for Fe_3C is smaller than for ferrite and are no longer distinguishable at the lowest resolutions.

Interestingly, both orientation and phase maps are practically identical for the as-collected data (512 × 512) and after a numerical binning down to 256 × 256 pixels. Even the embedded cementite particles are recognized despite the dominant diffracting signal from the ferrite matrix. This means that there is limited benefit from using a very large number of pixels. On the contrary, such results advocate for the use of a detector with large pixel sizes (i.e., higher intensities per pixel) rather than high resolutions. Notably, if large-pixel detectors (or binning) are used to reduce the exposure (and so experimental) time, there will be no substantial improvement in processing time with these lower resolution patterns, the latter being already highly optimized. Indeed, apart from the image processing time (e.g., background subtraction), the number of operations does not scale with the pattern resolution but with the subset m' of datapoints contained in the templates, as discussed in Section 2.1.

Pattern size: 512 × 512 256 × 256 128 × 128 64 × 64 32 × 32

Figure 2. Orientation and phase maps (Fe: red, Fe_3C: green) for a FeC tip computed with decreasing pattern resolution. The data with a phase reliability lower than 15 are disregarded.

4. Orientation Ambiguities

4.1. Detecting Ambiguities

One particular limitation of the ACOM/TEM approach that must be addressed is the famous 180° ambiguity problem. In transmission, the diffracting signal is limited to low Bragg angles, typically a few degrees. Consequently, the diffraction patterns may contain only the zero order Laue zone (ZOLZ) and the orientation determination is unsafe. A typical example is the {111} diffraction pattern for cubic materials whose ZOLZ exhibits a six-fold symmetry while the material has a three-fold symmetry. The absence of reflections belonging to a higher order Laue zone (HOLZ) prevents the crystal orientation from being distinguished between two possibilities that are related by a 180° rotation around the {111} ZA. This leads to possible misindexing and, consequently, damages the quality of the orientation maps and the reliability of the data.

A straightforward way to highlight orientation misindexing due to ambiguities is to draw the grain boundary map. Ambiguities will appear as spotty 'boundaries' within specific grains because misorientations related to the 180° rotation are higher than the threshold used to define a grain boundary (Figure 3a). Another characteristic signature of the induced error appears when the disorientations (minimum misorientation taking into account the symmetry of the crystal) measured along a line that crosses the area of interest is plotted: if the curve exhibits jumps of a few tens of degrees while the orientation is apparently unchanged (same color code) this is most probably related to a 180° ambiguity.

The capability of indexing software to handle such situations was questioned since the beginning of the development of ACOM/TEM approaches [14,15]. A specific workflow was developed to overcome the ambiguity issue and is described below.

(**a**) (**b**) (**c**)

Figure 3. Iron polycrystalline sample with numerous misindexings, (**a**) raw orientation map: colors are related to different orientations of the out-of-page *z*-axis; the superimposed black lines highlight grain boundaries defined as locations where the disorientation exceeds 5°; note the spotty aspect of dark blue and pink grains, (**b**) ambiguity map showing the safe areas (white) versus the ambiguous grains (black), (**c**) same as (**a**) but with orientations corrected through the procedure described in the main text.

4.2. Correcting Ambiguities

Highly symmetric diffraction patterns are intrinsically susceptible to ambiguities. If possible (single or multi-crystal), the sample should be tilted to avoid zone axes. If not (polycrystals), the missing information must be retrieved elsewhere. A possibility consists of extracting additional information from the local neighbourhood. Indeed, the orientation is never perfectly constant over a significant area: small orientation gradients are quite common because either the thin foil has a small curvature, or the material is intrinsically deformed. Consequently, a highly ambiguous spot pattern (i.e., highly symmetric) is surrounded by less problematic patterns. When detected, the ambiguity may be eliminated by extending the proper orientation from the most reliable points to the unreliable ones located in the close vicinity.

Therefore, a post-processing automated correction routine was developed. It consists of:

1. Calculating the local degree of ambiguity.
2. For ambiguous orientations, comparing the local orientation with that of neighbouring pixels.
3. Exchanging the orientation with the one corresponding to a 180° rotation around the closest zone axis. Of course, the 'corrected' orientation is kept only in the case of improvement (i.e., decrease of disorientation).

The degree of ambiguity is calculated through the template matching procedure: the selected template (Figure 4a) is tilted 180° around the ambiguity axis (Figure 4c) and the tilted template is compared to the acquired pattern (Figure 4b), producing a second value of the correlation index Q_2. The related indexes (Q_1 and Q_2) are compared to give the ambiguity parameter:

$$\text{Amb} = 100 * (1 - Q_2/Q_1) \quad (\text{safe} = 100, \text{unsafe} = 0), \tag{4}$$

Safe and unsafe grain may be noticed by plotting the ambiguity value in grey scale (Figure 3b). Notably, a rather confusing aspect of this construction is that a low value of the 'ambiguity' parameter means that the solution is highly ambiguous and vice-versa, i.e., it is perhaps better described as an 'unambiguity' parameter. The ambiguity axis around which this rotation should be performed is the closest ZA from the current orientation.

Practically, the orientation information is propagated from one pixel to the next, starting from the safest point with respect to ambiguities. Corrections are performed point by point with a random walking path that finally covers the whole area bounded by grain boundaries. As soon as a grain is completed, the process starts again for the remaining pixels, starting with the safest non-processed point, until the whole map is completed.

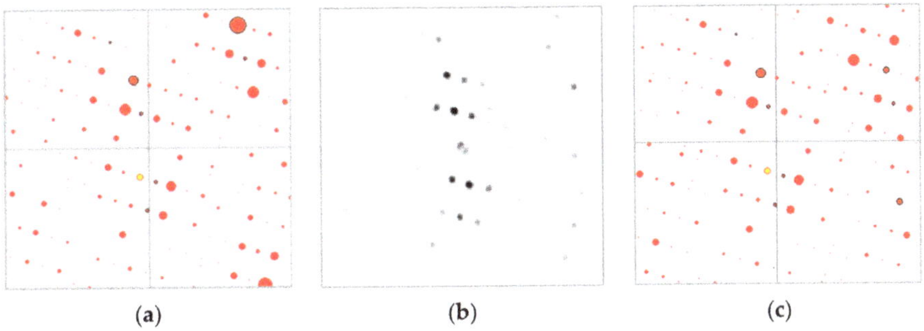

Figure 4. Illustration of the ambiguity parameter definition. In (**a**) the template orientation is ~[189], the ambiguity axis is [011] and the matching index Q_1 = 623 for the experimental pattern of NiSi shown in (**b**). In (**c**) the template orientation is ~[198] (i.e.,: ~[189] tilted 180° @ [011]) and the matching index Q_2 = 480 results in an ambiguity parameter Amb = 23.

Figure 3 compares the results before (Figure 3a) and after (Figure 3c) the ambiguity correction for an iron polycrystal. For body cubic centred structures, <111> and <112> ZA are the most problematic (respectively, blue and pink for the present colour code).

5. Retrieving Through-Thickness Information

In transmission, the electron beam is, by definition, crossing the sample and when the sizes of the structural features are smaller than the thin foil thickness, it collects information from all the overlapping grains and phases. A standard template matching indexing procedure extracts the signature of the dominant grains of phases so that the resulting maps are not simple cross sections of the sample, as is the case for EBSD tools, for example. This drawback comes with an attractive opportunity: for the through-thickness information being collected, it must be possible to deconvolute the combined signatures to identify all the superimposed elements. There are several attempts to reconstruct the 3D crystal nature by relating automated crystal orientation and tilt series mapping [16–18]. Such an approach will be described in the last section. But the 4D dataset acquired in a single scan already contains 3D features that may be exploited. This is performed in the following section that makes use of both correlation coefficient maps [19] and multi-indexing [20].

5.1. 3D Information Obtained from a Single Scan

With the template matching strategy, the acquired pattern is compared to all the pre-computed templates. The dominant one is retained but, in the case of overlapping crystals, additional candidates are also present. In order to recognize them properly, a good approach is to subtract the reflections related to the dominant grain before seeking for the next best match. The process, which may be repeated several times, is called multi-indexing and was described in detail elsewhere [20]. It may be used for orientation and phase identification [21,22].

The outcome of multi-indexing is a collection of all the orientations that were extracted either at the first or during the following indexations. The process is illustrated in Figure 5 for a copper nano device subjected three times to the multi-indexing procedure: it clearly shows that a given crystal is still detected after successive subtractions of the reflections related to the dominant grains.

A crystal exists if a non-negligible number of neighbor pixels have practically the same orientation. Consequently, the results are sorted into groups characterized by a limited orientation spread, typically 5°, and ranked by size (Figure 6a). As for a scree plot in principal component analysis, the smallest weights in the chart are related to noise or errors and thus the corresponding components are disregarded. As a side product of this procedure, the data is cleaned from misindexed components. The largest components are the constitutive crystals.

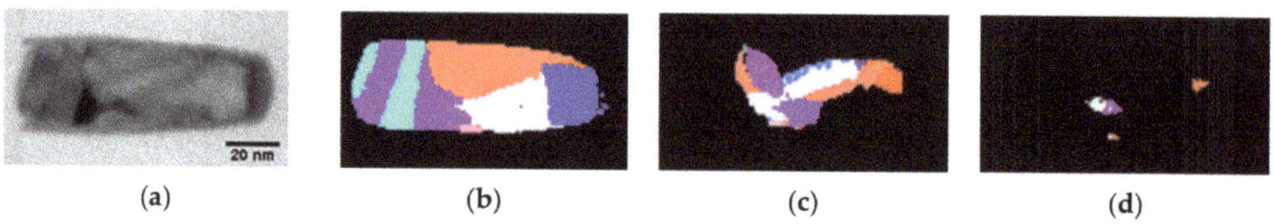

Figure 5. Cross section of a copper nano connector, (**a**) bright-field image, (**b**) first indexation, (**c**) second run, (**d**) third run. The grain highlighted in white extends as a non-dominant component after successive subtractions of the dominant grain reflections. The scan size is 110 × 60 pixels.

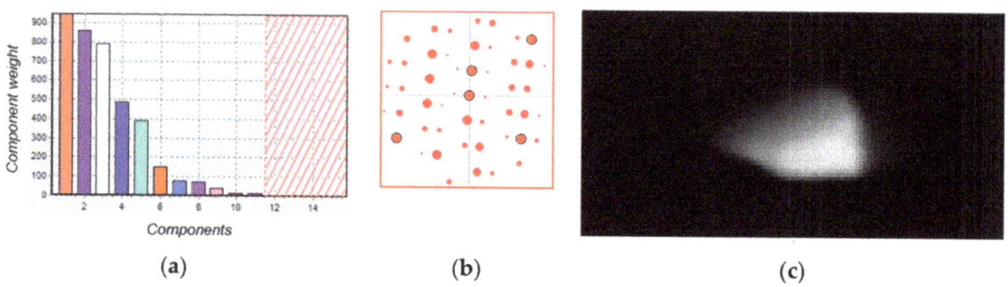

Figure 6. Frozen template VDF image construction: (**a**) the components are sorted by size and the smallest ones omitted, (**b**) each component is associated to a template, the one shown corresponds to the white bar in (**a**), the template is used as a multiple hole aperture to produce (**c**) a frozen template VDF image.

The grain orientation, defined either by an average over the component or by the orientation related to the highest correlation index for this particular grain, is associated to a unique template (Figure 6a). A second step consists of using this template as a mask for the construction of a combined virtual dark-field image [23], i.e., a numerical aperture that replicates the template is used to highlight the grain. The resulting 'frozen template virtual dark-field' (FTVDF) image reproduces the grain contour with, moreover, an intensity that is thickness dependent (Figure 6c).

The quality of the resulting image may be inferred by comparing it to the correlation coefficient map (CCM) [19,24]. The latter is a plot of the similarities between neighbouring diffraction patterns, calculated by a cross-correlation value, and highlights all material structural features that modify the diffracting conditions. The CCM looks similar to a projection of a semi-transparent sample and gives a direct view to 3D information, in particular the location and extent of grain boundaries (Figure 7a). The copper sample of Figure 5 contains numerous twins and the grain reconstructed in Figure 6c is bounded by an inclined twin boundary. As expected from the FTVDF image, the thickness of this crystal is at maximum in the lower part of the sample and vanishes in the upper part. This gradient is directly related to the inclined twin boundary shown in the CCM and illustrated by combining the two pictures in Figure 7b.

The same construction was used for all the components selected in Figure 6a. Of interest is the 'hidden' grain shown in Figure 7c that is clearly visible both in the CCM and in the FTVDF image but that was not detected through standard orientation mapping (a single indexing step, Figure 5b).

The capability to isolate grains from their surroundings provides a clear path towards 3D characterizations through tomographic reconstructions. The current state of such an attempt is described in the following section.

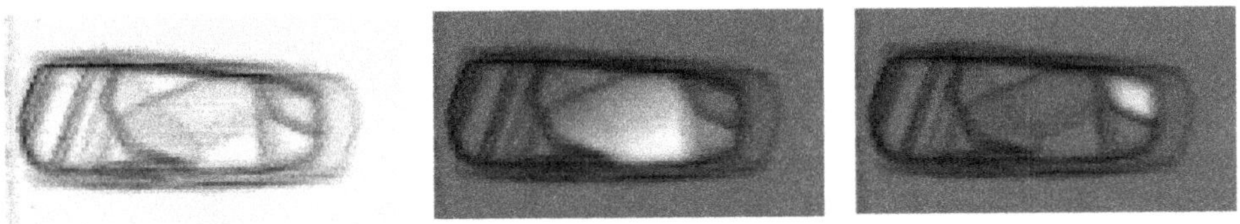

(a) (b) (c)

Figure 7. Three-dimensional information extracted from a single set of diffraction patterns: (**a**) the correlation coefficient map highlights inner boundaries, (**b**) a FTVDF image reproduces the through-thickness grain contour, (**c**) even for non-dominant components ('hidden' grain).

5.2. 3D Reconstruction from a Tilt Series

Despite the wealth of information contained within an orientation map, one of the main unknowns that remains is the complete 3D position of a grain within the sample. The in-plane (x, y) bounds of a grain are reasonably well defined, however, the z-position may be misinterpreted. It is common to believe that indexed grains within thin films are located at the same z-position, however when multiple grains exist within the projection this may not be the case. And whilst a better description of the volume of the grain can be calculated using FTVDF, as shown in the previous section, the z-ordering problem of overlapping grains still exists.

In order to more precisely determine the 3D position of grains within the sample, it is necessary to acquire images of the grains at different tilt angles (projections) within the TEM, a common practice in tomographic and stereographic imaging. Here, we will focus on tomographic reconstruction techniques from limited and restricted tilt experiments. Whilst the quality of tomographic reconstruction scales with the number of projections, from an experimental standpoint, this comes with the trade-off of experimental time on the TEM. The image acquisition time for direct imaging STEM (e.g., annular dark field) or TEM tomography experiments is typically on the order of seconds, whilst a single SPED dataset may take up to 30 min or more to acquire. This order of magnitude increase in frame acquisition time translates directly to experimental time. There may also be hardware-imposed limitations on the maximum achievable tilt angle. As a result, the microscope operator may wish to acquire datasets over a restricted tilt range, e.g., $\pm 30^\circ$, at large tilt steps, e.g., 10°.

Three-dimensional reconstruction from electron nano-diffraction was performed by multiple groups using different reconstruction techniques [16,25]. Eggeman et al. reported on crystallographic analysis of a Ni superalloy imaged by scanning precession electron tomography and used non-negative matrix factorization to isolate diffraction patterns from different grains [16]. Meng et al. used VDF reconstructions to create projections of grains within a TiN sample suitable for tomographic reconstruction [25]. Here, we utilise the indexed grain orientation to create FTVDF images of individual grains at different projection angles for tomographic reconstruction. The virtual apertures are calculated from grain components using the multi-indexing technique and tracked through the experimental tilt series enabling virtual reconstruction from their diffraction signature [26].

To achieve this, the components calculated from the orientation map of each tilt dataset are 'coupled' together, i.e., the rotation R that maps component orientations O from tilt dataset i is calculated such that $O_{i+1} = RO_i$. Then, by choosing an initial grain orientation from a given dataset, FTVDF images highlighting the grain in this dataset and all related tilt datasets can be produced, i.e., projections of the grain.

This methodology also produces an additional benefit: the projections of any individual grain may be used for tilt series alignment. Fiducial marker alignment methods [27] are well suited to data where the same feature can be tracked through the entire tilt series. These methods are also able to correct for stretching and shearing image artefacts that may occur due to the long acquisition time of a SPED dataset and sample drift in the TEM. In fact, the issue that often arises where overlapping features in the image obfuscate the tracked feature of interest may be completely avoided using FTVDF reconstructions, as the grain of interest can be unambiguously highlighted.

Tomographic reconstruction of an individual, aligned grain created through FTVDF can then proceed through established tomographic reconstruction techniques (e.g., back projection, SIRT [28]). An example reconstruction for the cold-drawn pearlitic steel tip, presented in Figure 2, using the methods described here is shown in Figure 8. In this case, each grain was reconstructed individually into a common volume, including the cementite grains described previously in Figure 2. The material processing conditions were described in [29]. The ferritic grains are predominantly columnar in shape and extend through the sample depth in one dimension. In contrast, the cementite particles are more spherical, and the reconstruction shows their location within the tip volume. Using this reconstruction method, we were able to individually reconstruct grains of different phases where the smallest grains were ~20 nm in one or more dimensions.

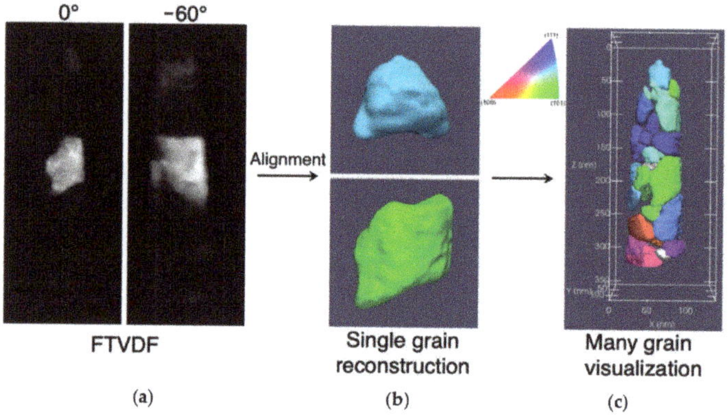

Figure 8. Three-dimensional reconstruction working process example for the FeC tip sample: (**a**) FTVDF reconstructed images of individual grains are used as projections for tomographic reconstruction, (**b**) an individual grain may be reconstructed, in this case using SIRT with 25 iterations. An isosurface rendering of each grain is presented here. (**c**) This process is applied to each grain to reconstruct the sample from its constituent grains. The grains are coloured by their orientation projected along z (colour map shown). Identified and reconstructed cementite particles are coloured white. For this reconstruction 9 projections were acquired over 360° with a minimum tilt step of 30°. The pixel size was 1.4 nm and each projection was 100 × 255 pixels. Renderings were performed using the Tomviz software [30].

6. Conclusions

Over the last fifteen years, the ACOM/TEM tool has proven as efficient and reliable towards the characterization of crystallographic orientations and phase distributions of crystallized materials, despite the physical constraints related to electron diffraction (dynamical effects, limited diffracting angles). The template matching algorithm detailed in this paper allows for rapid indexing and is effective even on low-resolution diffraction patterns. Numerous functionalities were implemented to improve the results; in particular, a dedicated functionality was implemented to solve the 180° ambiguity problem and a multi-indexing strategy was implemented to characterize the information within diffraction patterns superimposed by transmission of the electron beam through multiple grains. Moreover, the adaptability of the virtual reconstruction technique allows for the retrieval of some 3D spatial information from a single 2D dataset. By extending these techniques into 3D, e.g., scanning precession electron tomography, the full 3D description of grain shape and orientation may be calculated which creates the possibility to study grain relationships within the sample volume. This opens the door towards 3D characterization of complex nanomaterials and nano devices.

Author Contributions: Conceptualization, methodology, E.F.R., W.L. and M.H.; software, E.F.R. and P.H.; validation, E.F.R., P.H. and M.V.; formal analysis, E.F.R. and P.H.; investigation, E.F.R., X.Z. and P.H.; resources, E.F.R., M.H. and M.V.; writing—original draft preparation, E.F.R. and P.H.; writing—review and editing, E.F.R. and P.H.; visualization, E.F.R., X.Z. and P.H.; supervision, E.F.R., M.H. and M.V.; project administration, E.F.R.; funding acquisition, E.F.R., W.L., M.H. and M.V. All authors have read and agreed to the published version of the manuscript.

Funding: This research was partly funded by Agence National de la Recherche (ANR), grant number ANR-19-CE42-0017. M.H. acknowledges funding by the German Research Foundation (DFG) for funding via project HE 7225/11-1. The work has benefited from characterization equipment of the Grenoble INP-CMTC platform supported by the Centre of Excellence of Multifunctional Architectured Materials 'CEMAM' n°ANR-10-LABX-44-01 funded by the Investments for the Future programme.

Institutional Review Board Statement: Not applicable.

Informed Consent Statement: Not applicable.

Acknowledgments: We are grateful to Xuyang Zhou and Michael Herbig from MPIE at Düsseldorf (Germany) for their collaboration and particularly for supplying the steel raw data used to generate the results shown in Figures 2 and 8.

Conflicts of Interest: The authors declare no conflict of interest.

References

1. Rauch, E.F.; Véron, M. Automated Crystal Orientation and Phase Mapping in TEM. *Mater. Charact.* **2014**, *98*, 1–9. [CrossRef]
2. Rauch, E.F.; Portillo, J.; Nicolopoulos, S.; Bultreys, D.; Rouvimov, S.; Moeck, P. Automated Nanocrystal Orientation and Phase Mapping in the Transmission Electron Microscope on the Basis of Precession Electron Diffraction. *Z. Krist.* **2010**, *225*, 103–109. [CrossRef]
3. NanoMEGAS SPRL. Available online: https://nanomegas.com/ (accessed on 29 July 2021).
4. Mu, X.; Kobler, A.; Wang, D.; Chakravadhanula, V.S.K.; Schlabach, S.; Szabó, D.V.; Norby, P.; Kübel, C. Comprehensive Analysis of TEM Methods for $LiFePO_4/FePO_4$ Phase Mapping: Spectroscopic Techniques (EFTEM, STEM-EELS) and STEM Diffraction Techniques (ACOM-TEM). *Ultramicroscopy* **2016**, *170*, 10–18. [CrossRef]
5. Brunetti, G.; Robert, D.; Bayle-Guillemaud, P.; Rouvière, J.L.; Rauch, E.F.; Martin, J.F.; Colin, J.F.; Bertin, F.; Cayron, C. Confirmation of the Domino-Cascade Model by $LiFePO_4/FePO_4$ Precession Electron Diffraction. *Chem. Mater.* **2011**, *23*, 4515–4524. [CrossRef]
6. Guo, W.; Meng, Y.; Zhang, X.; Bedekar, V.; Bei, H.; Hyde, S.; Guo, Q.; Thompson, G.B.; Shivpuri, R.; Zuo, J.; et al. Extremely Hard Amorphous-Crystalline Hybrid Steel Surface Produced by Deformation Induced Cementite Amorphization. *Acta Mater.* **2018**, *152*, 107–118. [CrossRef]
7. Hart, M.J.; Bassiri, R.; Borisenko, K.B.; Véron, M.; Rauch, E.F.; Martin, I.W.; Rowan, S.; Fejer, M.M.; MacLaren, I. Medium Range Structural Order in Amorphous Tantala Spatially Resolved with Changes to Atomic Structure by Thermal Annealing. *J. Non-Cryst. Solids* **2016**, *438*, 10–17. [CrossRef]
8. Santiago, U.; Velázquez-Salazar, J.J.; Sanchez, J.E.; Ruiz-Zepeda, F.; Ortega, J.E.; Reyes-Gasga, J.; Bazán-Díaz, L.; Betancourt, I.; Rauch, E.F.; Veron, M.; et al. A Stable Multiply Twinned Decahedral Gold Nanoparticle with a Barrel-like Shape. *Surf. Sci.* **2016**, *644*, 80–85. [CrossRef]
9. Wang, Y.; He, J.; Mu, X.; Wang, D.; Zhang, B.; Shen, Y.; Lin, M.; Kübel, C.; Huang, Y.; Chen, H. Solution Growth of Ultralong Gold Nanohelices. *ACS Nano* **2017**, *11*, 5538–5546. [CrossRef]
10. Wang, A.; Leff, A.C.; Taheri, M.L.; Graef, M.D. A Statistical Dictionary Approach to Automated Orientation Determination from Precession Electron Diffraction Patterns. *Microsc. Microanal.* **2015**, *21*, 1247–1248. [CrossRef]
11. Nicolopoulos, S.; Bultreys, D.; Rauch, E. Precession Coupled Orientation/Phase Mapping on Nanomaterials with TEM Cs Microscopes. *Acta Crystallogr. A Found. Crystallogr.* **2012**, *68*, S104. [CrossRef]
12. Rauch, E.F.; Véron, M. Crystal Orientation Angular Resolution with Precession Electron Diffraction. *Microsc. Microanal.* **2016**, *22*, 500–501. [CrossRef]
13. Rauch, E.; Renou, G.; Veron, M. Reflection profile and angular resolution with Precession Electron Diffraction. In *European Microscopy Congress 2016: Proceedings*; Wiley-VCH Verlag GmbH & Co: Weinheim, Germany, 2016; pp. 665–666, ISBN 978-3-527-80846-5.
14. Wu, G.; Zaefferer, S. Advances in TEM Orientation Microscopy by Combination of Dark-Field Conical Scanning and Improved Image Matching. *Ultramicroscopy* **2009**, *109*, 1317–1325. [CrossRef]

15. Morawiec, A.; Bouzy, E. On the Reliability of Fully Automatic Indexing of Electron Diffraction Patterns Obtained in a Transmission Electron Microscope. *J. Appl. Cryst.* **2006**, *39*, 101–103. [CrossRef]
16. Eggeman, A.S.; Krakow, R.; Midgley, P.A. Scanning Precession Electron Tomography for Three-Dimensional Nanoscale Orientation Imaging and Crystallographic Analysis. *Nat. Commun.* **2015**, *6*, 7267. [CrossRef] [PubMed]
17. Kobler, A.; Kübel, C. Towards 3D Crystal Orientation Reconstruction Using Automated Crystal Orientation Mapping Transmission Electron Microscopy (ACOM-TEM). *Beilstein J. Nanotechnol.* **2018**, *9*, 602–607. [CrossRef] [PubMed]
18. Liu, H.H.; Schmidt, S.; Poulsen, H.F.; Godfrey, A.; Liu, Z.Q.; Sharon, J.A.; Huang, X. Three-Dimensional Orientation Mapping in the Transmission Electron Microscope. *Science* **2011**, *332*, 833–834. [CrossRef] [PubMed]
19. Kiss, Á.K.; Rauch, E.F.; Lábár, J.L. Highlighting Material Structure with Transmission Electron Diffraction Correlation Coefficient Maps. *Ultramicroscopy* **2016**, *163*, 31–37. [CrossRef]
20. Valery, A.; Rauch, E.F.; Clément, L.; Lorut, F. Retrieving Overlapping Crystals Information from TEM Nano-Beam Electron Diffraction Patterns: ACOM-TEM & OVERLAPPING CRYSTALS. *J. Microsc.* **2017**, *268*, 208–218. [CrossRef]
21. Rauch, E.F.; Véron, M. Methods for Orientation and Phase Identification of Nano-Sized Embedded Secondary Phase Particles by 4D Scanning Precession Electron Diffraction. *Acta Crystallogr. B Struct. Sci. Cryst. Eng. Mater.* **2019**, *75*, 505–511. [CrossRef]
22. Valery, A.; Rauch, E.F.; Pofelski, A.; Clement, L.; Lorut, F. Dealing with Multiple Grains in TEM Lamellae Thickness for Microstructure Analysis Using Scanning Precession Electron Diffraction. *Microsc. Microanal.* **2015**, *21*, 1243–1244. [CrossRef]
23. Rauch, E.F.; Véron, M. Virtual Dark-Field Images Reconstructed from Electron Diffraction Patterns. *Eur. Phys. J. Appl. Phys.* **2014**, *66*, 10701. [CrossRef]
24. Wang, S. Application of Diffraction Mapping on Crystal Grain Imaging. *Microsc. Microanal.* **2013**, *19*, 708–709. [CrossRef]
25. Meng, Y.; Zuo, J.-M. Three-Dimensional Nanostructure Determination from a Large Diffraction Data Set Recorded Using Scanning Electron Nanodiffraction. *IUCrJ* **2016**, *3*, 300–308. [CrossRef] [PubMed]
26. Harrison, P.; Zhou, X.; Das, S.M.; Viganò, N.; Lhuissier, P.; Herbig, M.; Ludwig, W.; Rauch, E. Reconstructing Grains in 3D through 4D Scanning Precession Electron Diffraction. *Microsc. Microanal.* **2021**, *27*, 2494–2495. [CrossRef]
27. Mastronarde, D.N. Fiducial Marker and Hybrid Alignment Methods for Single- and Double-axis Tomography. In *Electron Tomography*; Frank, J., Ed.; Springer: New York, NY, USA, 2006; pp. 163–185. ISBN 978-0-387-31234-7.
28. Goris, B.; Roelandts, T.; Batenburg, K.J.; Mezerji, H.H.; Bals, S. Advanced Reconstruction Algorithms for Electron Tomography: From Comparison to Combination. *Ultramicroscopy* **2013**, *127*, 40–47. [CrossRef] [PubMed]
29. Herbig, M.; Choi, P.; Raabe, D. Combining structural and chemical information at the nanometer scale by correlative transmission electron microscopy and atom probe tomography. *Ultramicroscopy* **2015**, *153*, 32–39. [CrossRef] [PubMed]
30. Hanwell, M.D.; Harris, C.J.; Genova, A.; Schwartz, J.; Jiang, Y.; Hovden, R. Tomviz: Open Source Platform Connecting Image Processing Pipelines to GPU Accelerated 3D Visualization. *Microsc. Microanal.* **2019**, *25*, 408–409. [CrossRef]

symmetry

MDPI

Article

Determination of Spinel Content in Cycled $Li_{1.2}Ni_{0.13}Mn_{0.54}Co_{0.13}O_2$ Using Three-Dimensional Electron Diffraction and Precession Electron Diffraction

Matthias Quintelier [1], Tyché Perkisas [2], Romy Poppe [1], Maria Batuk [1], Mylene Hendrickx [1] and Joke Hadermann [1,*]

[1] EMAT, Department of Physics, University of Antwerp, 2020 Antwerp, Belgium; matthias.quintelier@uantwerpen.be (M.Q.); romy.poppe@uantwerpen.be (R.P.); maria.batuk@uantwerpen.be (M.B.); Mylene.hendrickx@uantwerpen.be (M.H.)

[2] Department of Engineering Management, Faculty of Business and Economics, University of Antwerp, 2000 Antwerp, Belgium; tyche.perkisas@uantwerpen.be

* Correspondence: joke.hadermann@uantwerpen.be

Abstract: Among lithium battery cathode materials, $Li_{1.2}Ni_{0.13}Mn_{0.54}Co_{0.13}O_2$ (LR-NMC) has a high theoretical capacity, but suffers from voltage and capacity fade during cycling. This is partially ascribed to transition metal cation migration, which involves the local transformation of the honeycomb layered structure to spinel-like nano-domains. Determination of the honeycomb layered/spinel phase ratio from powder X-ray diffraction data is hindered by the nanoscale of the functional material and the domains, diverse types of twinning, stacking faults, and the possible presence of the rock salt phase. Determining the phase ratio from transmission electron microscopy imaging can only be done for thin regions near the surfaces of the crystals, and the intense beam that is needed for imaging induces the same transformation to spinel as cycling does. In this article, it is demonstrated that the low electron dose sufficient for electron diffraction allows the collection of data without inducing a phase transformation. Using calculated electron diffraction patterns, we demonstrate that it is possible to determine the volume ratio of the different phases in the particles using a pair-wise comparison of the intensities of the reflections. Using this method, the volume ratio of spinel structure to honeycomb layered structure is determined for a submicron sized crystal from experimental three-dimensional electron diffraction (3D ED) and precession electron diffraction (PED) data. Both twinning and the possible presence of the rock salt phase are taken into account. After 150 charge–discharge cycles, 4% of the volume in LR-NMC particles was transformed irreversibly from the honeycomb layered structure to the spinel structure. The proposed method would be applicable to other multi-phase materials as well.

Keywords: 3D ED; PED; NMC; phase ratio; battery

1. Introduction

Battery cathode materials based on $Li_{1+\delta}(Ni_{1-x-y}Mn_xCo_y)_{1+\delta}O_2$ are widely investigated in various compositions for the different advantages that specific Li:Ni:Mn:Co ratios can give in capacity and power density. Voltage fade and capacity fade are not yet optimally subdued, partially due to the creation of secondary phases during cycling [1–4]. Their parent structure is a layered $LiCoO_2$ type, and in the case of lithium-rich compounds ($\delta > 0$)—such as the $Li_{1.2}Ni_{0.13}Mn_{0.54}Co_{0.13}O_2$ (LR-NMC) studied in this paper—there is a honeycomb order between the lithium atoms and the transition metal atoms in the transition metal layers (Figure 1a). Upon cycling, some of the transition metal atoms migrate into the octahedral positions within the lithium layers via the tetrahedral positions, which can occur in an ordered manner, creating a spinel structure locally [1,5] (Figure 1c). There can also be a further transformation to a rock salt type structure with disorder between the lithium atoms and the transition metal atoms [6] (Figure 1d). Furthermore, surface layers with different compositions, both amorphous and crystalline, can be formed [7].

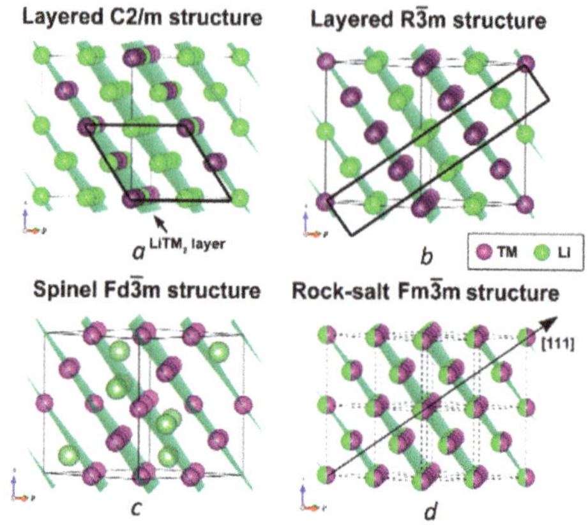

Figure 1. Crystal structure of $Li_{1.2}Ni_{0.13}Mn_{0.54}Co_{0.13}O_2$ (LR-NMC). In pristine LR-NMC, (**a**) there is a honeycomb order between the lithium atoms and the transition metal atoms in the transition metal layers, compared to the layered $LiCoO_2$ type structure in (**b**). During cycling, a spinel type structure (**c**) and a rock salt type structure (**d**) can be formed. The green and purple atoms represent Li and transition metals (TM = Ni, Mn, Co), respectively. For clarity, oxygen atoms are not shown.

The complex formation of secondary phases upon cycling is easiest to investigate by transmission electron microscopy (TEM), as this allows the direction observation of the structure at atomic resolution using different high-resolution imaging techniques [4,7,8]. However, it has been demonstrated that these compounds are sensitive to the electron beam, and imaging these particles has the same effect of turning the layered structure into a spinel structure [8–10]. Because of this, it is difficult to conclude that a certain change in the structure is due to cycling and not due to the electron beam. Moreover, it has been shown that twins originating from $Fm\bar{3}m$ rock salt type structure to $R\bar{3}m$ layered structure (Figure 1d,b) symmetry lowering during synthesis can easily be mistaken for the occurrence of spinel or rock salt phases when relying only on images [11–13].

From TEM imaging, the spinel structure is observed at the surface of the particles [1]. In situ synchrotron X-ray diffraction showed that there are also transition metals in the octahedral positions of the lithium layers (corresponding to transition metal positions in spinel but not in layered phases) in the bulk of the particles [2]. X-ray powder diffraction needs modelling and the implementation of various assumptions [2]. However, single-crystal X-ray diffraction cannot be used to determine the evolution of the structure because the lithium diffusion paths are too long to give a representative behavior in single crystals of a large enough size for single crystal X-ray diffraction. TEM imaging techniques are inadequate to detect the occurrence of spinel in the bulk of the crystals because the thickness of the crystals prevents the obtaining of clear images of bulk areas.

Therefore, we suggest using electron diffraction instead. In contrast to single-crystal X-ray or neutron diffraction, single-crystal electron diffraction can be done on submicron-sized crystals, which is the size used in battery cathode materials. When compared to TEM imaging, electron diffraction can readily distinguish whether extra reflections are due to twinning or due to the formation of the spinel phase [11–13]. Furthermore, electron diffraction requires lower beam intensities while still giving access to information at the level of atom coordinates, even of the lithium ions [14–18].

Still, whereas the amount of transition metal atoms in tetrahedral positions can be obtained using least squares refinement from three-dimensional electron diffraction (3D ED) data for Ni-rich $LiNi_{1-x-y}Mn_xCo_yO_2$ compounds [19], least squares refinement of LR-NMC from 3D ED data has not yet been successful due to the presence of strong diffuse streaks along the c* axis in the electron diffraction patterns. The intensity maxima along the diffuse streaks are caused by twins originating from extra $R\bar{3}m$ (Figure 1b) to

$C2/m$ (Figure 1a) symmetry lowering. The diffuse streaked character of these reflections is due to the nanoscale stacking of slabs of the different twins. When the samples are cycled, the occurrence of spinel and rock salt phases further complicates the refinement. Refining the phase ratio of spinel to honeycomb layered structure from the intensity profile of the diffuse streaks [20–23] is not a possibility for LR-NMC, since the spinel and honeycomb layered characteristic reflections form separate rows.

In this paper, we demonstrate the difference in sensitivity to the electron beam between pristine and charged LR-NMC type materials, demonstrate that it is possible to obtain electron diffraction patterns without introducing extra spinel, and provide a strategy to determine the ratio between the different phases present in a particle from 3D ED when least squares refinement does not give a solution due to a complex microstructure. Using this method, we determine the percentage spinel present in the volume of a submicron sized LR-NMC crystal in the discharged state after it has been cycled 150 times.

2. Materials and Methods

The details on the synthesis and cycling of $Li_{1.2}Ni_{0.13}Mn_{0.54}Co_{0.13}O_2$ (LR-NMC) and $Li_{1.2}Ni_{0.13}Mn_{0.513}Co_{0.13}Sn_{0.027}O_2$ (Sn-doped LR-NMC) are given in reference [24]. Both materials were galvanostatically cycled between 3.0 and 4.55 V vs. Li/Li+ at C-rates of C/20 during the initial two cycles and C/10 during the subsequent cycles. The effect of the electron beam was tested on both pristine and one time charged Sn-doped LR-NMC. For the application of the phase ratio determination method explained in this paper, a 150-times cycled LR-NMC sample was used in the discharged state.

All electron diffraction patterns were taken on Tecnai transmission electron microscope (Thermo Fisher, Eindhoven, The Netherlands) operated at 200 kV. The experiments with charged and discharged particles were done using a vacuum transfer holder (Gatan, Pleasanton, CA, USA) to prevent exposing the particles to air; the pristine sample was studied using a conventional double tilt holder (Gatan, Pleasanton, CA, USA).

In-zone electron diffraction patterns were acquired with precession electron diffraction (PED) using a Digistar precession attachment (Nanomegas, Brussels, Belgium), with a precession angle of 1°. Three-dimensional electron diffraction (3D ED) data were acquired with the manual electron diffraction tomography method, at intervals of 1° tilt. All electron diffraction patterns were recorded using an FEI Eagle 2k CCD camera (Thermo Fisher, Eindhoven, The Netherlands).

For validation of the method, in-zone electron diffraction patterns of core-shell particles were calculated using DISCUS (version 6.00.02) [25,26].

The reconstruction of the reciprocal space sections and the extraction of the intensities of the reflections were done using PETS (version 2.0) [27,28]. For the experimental data, the intensities were integrated using the fit profile option. For the data calculated in DISCUS, the intensities were integrated using the maximum intensity option. The fit profile option applies corrections, such that the resulting intensities I in the (hkl, I)-lists agree with the formula $I(hkl) = F(hkl)F^*(hkl)$—with $F(hkl)$, the structure factor for electron scattering, and $F^*(hkl)$, its complex conjugate. The maximum intensity option does not apply corrections. As DISCUS calculates the intensities using $I(hkl) = F(hkl)F^*(hkl)$, no corrections should be applied for the analysis of the calculated data.

For the analysis of the intensity ratios from the (hkl, I)-lists, an in-house written Matlab code "Crystalblender" was used, which is available upon request. Refinements were done using JANA2006 (version 25/10/2015) [29]. Fitting of the histograms to acquire the mean value and standard deviation were done using Fityk (version 1.3.1) [30].

3. Results and Discussion

3.1. Effect of the Electron Beam on the Spinel Ratio in LR-NMC Battery Cathode Materials When Using Electron Diffraction

While the effect of the electron beam on the spinel content in $Li_{1.2}Ni_{0.13}Mn_{0.54}Co_{0.13}O_2$ (LR-NMC)-type materials has been well documented for TEM imaging techniques [8–10], no literature is available on the effect of electron diffraction on the amount of spinel. Electron diffraction can be performed with a good signal-to-noise ratio using much lower electron doses than TEM imaging [18,31–33]. Therefore, we first investigated the effect of the electron beam in electron diffraction setting using a pristine and a charged sample of the same material at different beam intensities. For this experiment, we used $Li_{1.2}Ni_{0.13}Mn_{0.513}Co_{0.13}Sn_{0.027}O_2$ (Sn-doped LR-NMC), which is a typical doped LR-NMC compound that rapidly transforms to the spinel phase under the electron beam conditions used for imaging [24]. If the particles can be illuminated by electron diffraction at appropriate intensities for a long enough time without the creation of the spinel phase, three-dimensional electron diffraction (3D ED) data can deliver reliable information on the change in spinel content due to charging and discharging.

To make a first visual evaluation during the ongoing experiment, we chose to observe $[010]_H = [0\bar{1}1]_{SP}$ (H—honeycomb layered phase, SP—spinel phase) in-zone electron diffraction patterns during irradiation (See Figure S1 for the relations between the different in-zone patterns based upon the transformation matrices described in Section 3.2). This zone includes lattice points, where reflections only occur in the presence of the spinel phase. We used particles that were almost in-zone and thus required only minimal orientation, to avoid as much irradiation as possible before the first in-zone pattern could be recorded. Since the particles did not show pronounced anisotropy, a small diameter should be associated with a small thickness. Therefore, we chose particles with a small diameter, around maximally 200 nm in diameter, to keep the contribution of dynamical scattering low. For typical lithium battery cathode materials, the kinematical approach in treating 3D ED or precession electron diffraction (PED) data is adequate, due to the low atomic number of the elements (inherent to battery requirements to keep the portable batteries as light as possible) [16,31–34]. A selected area aperture of 500 nm was used, which allowed the particles to fit completely in the aperture with some room for movement while still maintaining a sufficient quality signal-to-noise ratio, with as low as possible background from both the central beam and from diffraction on the amorphous carbon support. PED patterns were taken directly after in-zone orientation was obtained (Figure 2a,c). The particles were then irradiated using a beam with 45(14) $e^-/Å^2s$—similar to the intensity we normally use for 3D ED, where no special care needs to be taken concerning beam sensitivity, and similar to what is used for conventional selected area electron diffraction. PED patterns were then acquired every 5 min for 60 min. We saw no changes for either the pristine or the charged particles (Figure 2b,d). For the charged particles we then focused the electron beam to obtain a high intensity ($1.1(4).10^4$ $e^-/Å^2s$). The focused beam had the size of the particle, to ensure that the entire particle was homogeneously irradiated. During this irradiation, new PED patterns were taken every 5 min. Only with this higher electron dose rate did the spinel-only reflections appear after 20 min (Figure 2e) and keep increasing in intensity (Figure 2f). Thus, our data showed it was possible to acquire the necessary dataset for the determination of the spinel content present in the crystal without observably affecting it during at least 60 min, using normal electron diffraction settings. A manually obtained 3D ED series takes less than 60 min, and automated ones take only a few minutes [35–39], while an in zone PED pattern takes mainly the time necessary to orient the pattern. Therefore, 3D ED series and in-zone PED patterns can be obtained without affecting the spinel content. Note that for these LR-NMC materials it is not even necessary to use conditions such as those used for beam sensitive materials, as used for example by Palatinus et al. for hydrogen-containing materials [40], Kolb et al. for organics [41] or by Kodjikian et al. [42]—although this would also still be a possibility for ensuring a lower beam effect if deemed necessary.

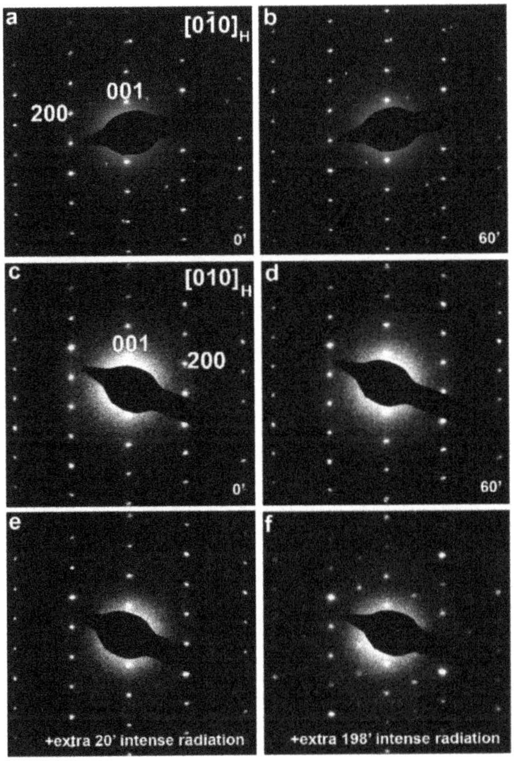

Figure 2. Selected area precession electron diffraction (PED) patterns of the $[0\bar{1}0]_H$ zone of a pristine particle after (**a**) 0 min and (**b**) 60 min at electron dose rate of 45(14) e/Å^{2}s, showing no change. PED patterns of the $[010]_H$ zone of a one-time charged particle after (**c**) 0 min and (**d**) 60 min at electron dose rate of 45(14) e^-/Å^{2}s, showing no change; (**e**) after an extra 20 min at electron dose rate 1.1(4).10^4 e^-/Å^2 s, weak reflections belonging to the spinel phase have appeared; (**f**) after 198 extra minutes at electron dose rate 1.1(4).10^4 e^-/Å^2 s, showing very clear reflections from the spinel phase. Slight differences in background intensity are due to adjustments of the total grey level interval for best visualization of all reflections.

3.2. Determination of the Amount of Spinel in a LR-NMC Submicron-Sized Crystal

Whereas we showed in the previous section that electron diffraction gives a reliable representation of the amount of spinel in LR-NMC crystals, quantification of the amount of spinel is far from routine. As explained in the introduction, the multitude of defects and phases in LR-NMC has so far prevented a refinement of the phase ratio or the occupancy of the positions from electron diffraction single crystal data. Refinements of the occupancies have been performed using powder X-ray diffraction, but need extensive modelling and assumptions [2].

Therefore, we propose to use the pair-wise intensity ratios of the reflections in electron diffraction data to obtain the phase ratio. For most reflection pairs, a certain intensity ratio will correspond to a unique phase ratio. Therefore, the phase ratio can be obtained from the intensity ratio. If such analysis was done based on conventional in-zone electron diffraction patterns, there would be several sources of errors—most importantly deviations in the intensities due to slight misorientation and deviations due to dynamical scattering. Therefore, we will use 3D ED data, for which the reflection intensities are integrated over several out-of-zone frames, which considerably reduces the dynamical contributions, and in-zone PED data, where the precession reduces dynamical scattering.

We only studied particles that fit entirely within the 200 nm selected area aperture as a criterium of their size, to further limit the dynamical scattering contribution to the intensities. The phase ratio is calculated by comparing the experimental intensity ratio with the theoretical intensity ratio curve calculated for varying spinel content. Each pair thus results in a contribution to the distribution of the experimental spinel content. The standard deviation σ_I of the intensities is taken into account by calculating the standard deviation of the experimentally determined spinel content using error propagation and building a normal distribution, with μ being the mean spinel content and σ its standard deviation. Thus, instead of comparing one value (the experimentally determined intensity ratio) with the (spinel content, intensity ratio) curve, an entire distribution is compared, resulting in a likewise distribution for the spinel content.

For obtaining a single (*hkl*, *I*)-list containing all reflections of both the spinel and the honeycomb layered phase, both unit cells need to be transformed into a common supercell. A similar procedure was used for indexing twins in 3D ED data [43].The supercell is only used to allow indexing of all reflections, after which the transformation matrices are used to calculate the structure factor per volume unit using the corresponding reflections in the original unit cells. This allows for the user to use the original cell's CIF files and not spend time developing CIF files for the supercell. The intensities are then calculated corresponding to a continuously changing phase ratio by parametrizing the structure factor using the phase's volume percentage x as:

$$\frac{F}{V} = x\frac{F_A}{V_A} + (1-x)\frac{F_B}{V_B} \tag{1}$$

with $F_{A(B)}$ the structure factor originating from phase $A(B)$, and $V_{A(B)}$ the volume of the respective unit cells $A(B)$, which leads to an intensity:

$$I = k\left|x\frac{F_A}{V_A} + (1-x)\frac{F_B}{V_B}\right|^2 .V^2, \tag{2}$$

with k a scale factor depending on the thickness of the crystal and the beam intensity. The scale factor will be eliminated when taking the ratio of the intensities. Details on the derivation of this formula can be found in the Supplementary Materials. The relation of the basis vectors of the common supercell to the basis vectors of the honeycomb layered (H) unit cell and the spinel (SP) unit cell is indicated in Figure 3. The transformation matrices for both phases to this common supercell are:

$$P_{H\rightarrow super} = \begin{pmatrix} 1 & 1 & 1 \\ -1 & 1 & 0 \\ 0 & 0 & 3 \end{pmatrix} \tag{3}$$

$$P_{SP\rightarrow super} = \begin{pmatrix} 0.5 & 0.5 & -1 \\ 0.5 & -1 & -1 \\ -1 & 0.5 & -1 \end{pmatrix} \tag{4}$$

Note that this common supercell also allows the indexing of all rock salt reflections, since these overlap with spinel and honeycomb layered reflections, as the rock salt structure is the parent subcell of the honeycomb layered structure and the spinel structure (Figure 1). For clarity and comprehension, several calculated main zone electron diffraction patterns are included in Figure S1, showing next to each other the indices in honeycomb layered, spinel and rock salt structures, and in the common supercell. For the remainder of this paper, all indices will be given relative to the supercell axes, as indicated by the subscript S.

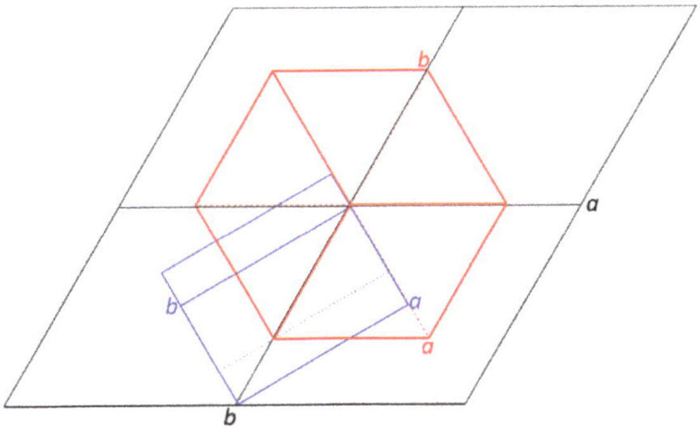

Figure 3. Relation between the basis vectors of the honeycomb layered structure (blue), the spinel structure (red) and the common supercell (black), projected along the c-axis of the supercell. The rock salt structure axes are the halved spinel axes. The labels a and b, indicating the basis vectors, are placed at the ends of the basis vectors, the origin is at the center of the scheme.

To test the approach, we used in-zone electron diffraction patterns calculated in DISCUS. Spherical core-shell particles with a honeycomb layered (*C*2/*m*) core [44] (ICSD 237940) and a spinel ($Fd\bar{3}m$) [45] (ICSD 40485) shell were modelled with varying ratios. For reasons of feasibility, no stacking faults or twinning were included in the model. All transition metals (Z_{Mn} = 25, Z_{Co} = 27, Z_{Ni} = 28) were approximated as Mn for the DISCUS calculations to avoid artefacts from the short-range transition metal order. To build a core-shell model in DISCUS, the cell parameters for both phases need to be identical. This was achieved by transforming both structures to the same common supercell as we used for the indexation of the experimental 3D ED series. Figure 4 shows the cross-section of a spherical crystal with 10 × 10 × 10 supercell unit cells and a spinel/honeycomb layered ratio of 50:50.

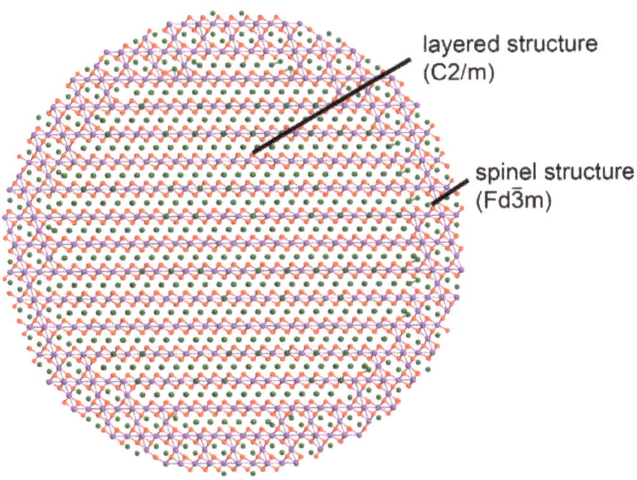

Figure 4. Cross-section of a spherical core-shell LR-NMC particle used for calculating electron diffraction patterns using DISCUS. The particle has a size of 10 × 10 × 10 supercell unit cells with a honeycomb layered core and a spinel shell. The spinel/honeycomb layered ratio used for the simulation of this particle is 50:50. Green, purple and red spheres represent lithium, transition metals and oxygen, respectively.

The $[\bar{1}10]_S$ (and $[100]_S$) electron diffraction patterns were calculated in DISCUS for a spinel content of 2, 5, 10, 20 and 50% (examples shown in Figure 5; Figure S2 for $[100]_S$). These patterns can be compared to Figure S1 for the origin of each reflection. Subsequently, the intensities of the reflections were extracted using PETS2 software [28] to get an (*hkl*, *I*)-list. Figure 6 shows the lattice points where PETS2 extracted the intensities using the supercell.

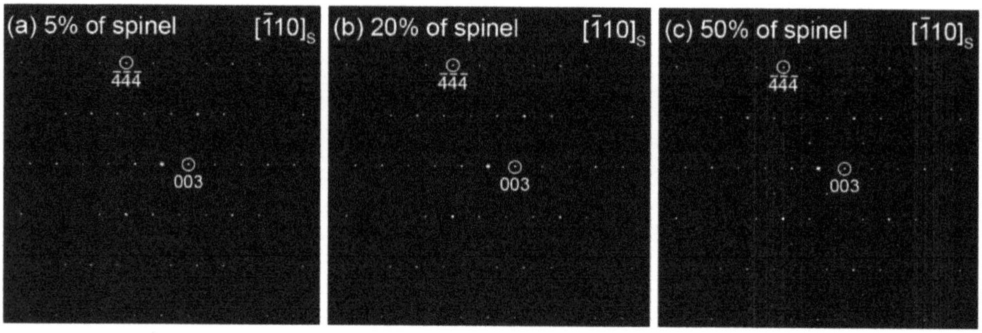

Figure 5. $[\bar{1}10]_S$ electron diffraction patterns calculated for, respectively, 5%, 20% and 50% of spinel, for a spherical crystal with $10 \times 10 \times 10$ supercell unit cells.

We tested different sets of reflection pairs: purely spinel over purely honeycomb layered, purely spinel over combined spinel and honeycomb layered, combined spinel and honeycomb layered over purely honeycomb layered, and combined spinel and honeycomb layered for both reflections (Figure S3). Only the pairs where one reflection was purely spinel and the other had a contribution from both the honeycomb layered and the spinel structure resulted in the correct spinel content. Using such reflection pairs, the (spinel content, intensity ratio) curve always shows a monotonous behavior that can be used for pinpointing which phase ratio corresponds to a certain intensity ratio. Reflection pairs in which both reflections have contributions from both phases often have multiple solutions for the spinel content, while for the other failing combinations, presumably, the effect of variations in spinel content is too small for reliable spinel content determination.

Figure 6 shows the (spinel content, intensity ratio) curve for an example reflection pair $\bar{1}\bar{1}5$:006, where $\bar{1}\bar{1}5$ is purely contributed by the spinel structure and 006 is a reflection common to both the spinel structure and the honeycomb layered structure. There is a good agreement of the curve obtained using the proposed equation $I = | x \frac{F_A}{V_A} + (1-x)\frac{F_B}{V_B} |^2$ to the discrete points obtained from Discus calculations for the core-shell particle.

The histograms corresponding to the datasets for the different spinel contents are shown in Figure 7, in which the frequency is the frequency of the corresponding x values where each reflection pair's intensity ratio can be represented with a distribution (taking its standard deviation into account). Drawing randomly from this distribution (weighing with their respective likelihood) thus gives rise to a distribution in x, as every one of the drawn values is mapped unto its corresponding x value. It was found that 100 drawings from the intensity ratio distribution was enough to represent it accurately—hence the factor of 100 between frequency and number of considered pairs. A clear peak is visible for the ratios of 2, 5, 10 and 20%, located at resp. 0.021(6), 0.05(1), 0.10(2) and 0.17(3). For the 50% spinel ratio, there are two peaks—the reason for this remains elusive to us—at 0.29(3) and 0.54(5). The numbers in the brackets are the half-width-half-maximum of the peaks. Thus, the proposed method gives reliable results for up to 20% spinel, which is well suited for the expected amount of spinel based on synchrotron X-ray data [2].

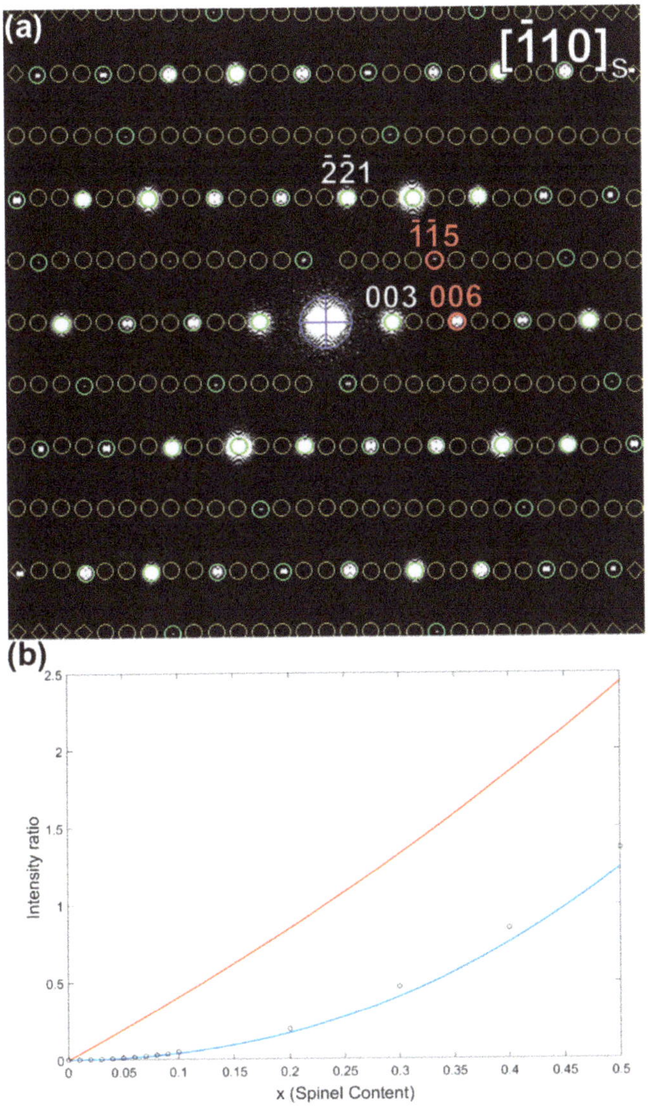

Figure 6. (**a**) Calculated $[\bar{1}10]_S$ zone for a spherical honeycomb layered–spinel core-shell particle with 2% of spinel. The circles indicate the positions of all reciprocal lattice points corresponding to the supercell ($I > 3\sigma$—green, $I < 3\sigma$—yellow). Red circles indicate the reflections used for the graph in (**b**), which shows as an example the intensity ratio of the $\overline{11}5$ to the 006 reflection versus the spinel content x (in percentage) as calculated in Discus (discrete dots), using Crystalblender with the formula $I = \mid x \frac{F_A}{V_A} + (1-x)\frac{F_B}{V_B} \mid^2$ implemented in this paper (blue line), and for comparison with the alternative formula $I = (x)I_A + (1-x)I_B$ (red line).

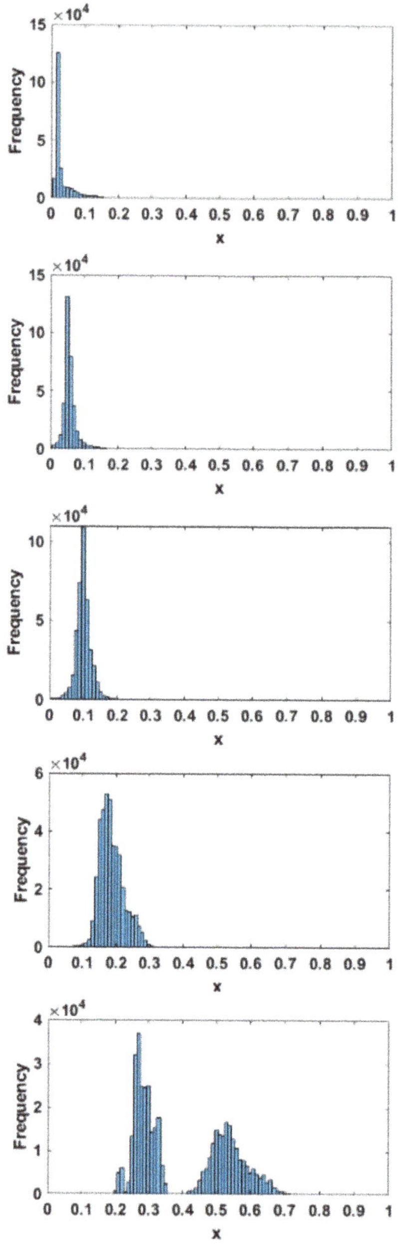

Figure 7. Histogram plots obtained using Crystalblender for the calculated DISCUS $[\bar{1}10]_S$ patterns with ratios of, from top to bottom, 2, 5, 10, 20 and 50% spinel. For these intensity ratios, the reflection pairs considered are all combinations of only reflections that only have $|F_A| > 0.0005$, $|F_B| < 0.0005$ for the numerator, and only reflections with $|F_A|$, $|F_A| > 0.0005$ for the denominator, and with h, k, $l < 15$. The frequency refers to the frequency of the x-values when the ratios are represented by 100 drawings from distributions of the intensity ratios with their standard deviations. The number of contributing pairs is the frequency divided by 100, as explained in the main text.

The calculated patterns are from idealized particles without twins, stacking faults, or rock salt structure contributions. This allows us to also attempt a conventional refinement to compare the results. The same (*hkl*, *I*)-lists were used as input for the refinement using JANA2006 [29]. We first attempted a two-phase least squares refinement with all structural parameters fixed to the literature values for the two phases, only refining the amount of the different phases and the scale factors. The refinement converged to incorrect, lower spinel contents (e.g., 1.1% instead of 10% of spinel, 4.9% instead of 20% of spinel and 32.8% instead of 50% of spinel, Table S1). An explanation for this can be found in the assumption in the least squares refinement that the intensities are due to separate crystals of the two phases, with the intensities being the sum of the separate intensities. However, in the core-shell particles, modelled after real observations made for LR-NMC cycled particles, both phases are present in the same particle. In our pair-wise intensity ratio approach, we have translated this into a contribution of both phases into the structure factor (Equation (5)) instead of a contribution only at the level of the intensities (Equation (6)). As described in Section 3.2, this results in intensities of supercell reflections as according to:

$$I = \left| \, x \frac{F_A}{V_A} + (1-x) \frac{F_B}{V_B} \right|^2 \tag{5}$$

instead of:

$$I = (x)I_A + (1-x)I_B, \tag{6}$$

which gives a resulting combined intensity *I* that is closer to that of a core-shell particle. As can be seen on Figure 6b, using Equation (6; red line) would result in lower apparent spinel contents than Equation (5; blue line) for the same intensity ratios.

Next, we attempted to refine the phase ratio using mixed occupancies in the supercell. This was not successful, as the number of refinable parameters was too high for the small number of reflections present in the single in-zone patterns. As a result, the refinement did not converge. Since this method assumes mixing of the phases at the structure factor level, we would expect that the least squares refinement of the occupancies would result in the correct spinel content. Unfortunately, simulating a full 3D ED series with the core-shell model in DISCUS to have sufficient reflections for the number of refinable parameters was not feasible.

Finally, the same ratio used as input could be successfully retrieved (although not by a real "refinement") by comparing the R factors for the agreement of the intensities for different fixed occupancies, leaving only the scale factors as refinable. Supercell models for 0–100% of spinel content with steps of 1% were created and tested versus the intensities obtained from the calculated diffraction patterns. This allowed us to pinpoint the correct phase ratio, as shown for example in Figure 8. In the case of the $[100]_S$ zone, the determined ratio was correct for 2, 5 and 10% of spinel, whereas for 20 and 50% the refined ratio was smaller than the correct one. Compared with the method based on pair-wise intensities, this phase ratio determination using the R factors demanded considerably more time and handling.

Thus, for calculated data, the pair-wise comparison of intensities and a search for the lowest R-factor using fixed-ratio input models gave comparable and correct results for the spinel contents of the two-phase honeycomb layered-spinel crystals up to, respectively, 20% and 10%, whereas two-phase least squares refinement with fixed structures gave too-low spinel contents.

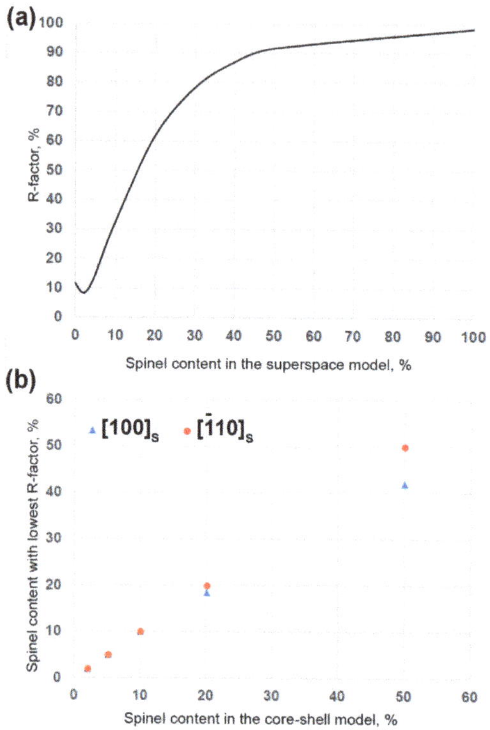

Figure 8. Results for least squares refinement using models with fixed ratios. (**a**) R factors obtained for the 2% spinel model. (**b**) Spinel contents with lowest R factor for the different $[100]_S$ and $[\bar{1}10]_S$ patterns with different spinel contents simulated in DISCUS.

3.3. Application to Experimental Data

We applied the phase ratio determination by pair-wise comparison of intensities on both an in-zone $[100]_S$ PED pattern and a 3D ED series acquired on the same LR-NMC crystal that was 150 times cycled and left in the discharged state (Figure 9). In contrast to the diffraction patterns calculated from the core-shell particle in the previous section, cycled LR-NMC crystals might also contain the rock salt phase. Therefore, all (*hkl*, *I*)-lists were filtered in Excel to remove all reflections for which all *h*, *k* and *l* are even and simultaneously $h + k + 1 = 6n$ in the supercell indexation, which covers all rock salt structure reflections (this can be easily verified on Figure S1). We cannot remove possible residual rock salt contributions present in the other reflections by multiple scattering, and we assume this effect is negligible. Besides the rock salt phase, the cycled LR-NMC crystals we studied contained stacking faults and twins caused by $R\bar{3}m$ to $C2/m$ symmetry lowering (extra details on the twinning in this material can be found in the Supplementary Materials). In such cases, each reflection that is due to only the layered phase in the (*hkl*, *I*)-lists will always have a contribution from only one twin, while the reflections overlapping with spinel contain contributions from all twins together. The intensities contributed by the different twins to the listed *hkl* will be different because the actual *hkl* for those twins at those positions will differ. Therefore, we expanded the calculation to take into account those different intensities by splitting up the layered intensities to consist of 1/3 of each twin. The estimate of 1/3 is based on a visual interpretation of the [100] PED pattern and *0kl* section of the 3D ED data. Additionally, thermodynamically it is likely that the three twins occur in equal amounts. The results with and without taking the twinning into account are shown in Figure 10. Not taking the twins into account results in a histogram with multiple peaks (Figure 10), while introducing the twins results in a clear single peak with only slight signs of a shoulder and tail. The latter are probably due to residual contributions from the two other twins if our equal distribution was not perfectly valid, as well as due to residual dynamical scattering. Dynamical scattering will redistribute the intensity from the strong reflections to the weaker reflections. In electron diffraction patterns of LR-NMC, the strongest reflections are the reflections common to both honeycomb layered and spinel, while the reflections purely from the spinel structure are among the weakest. Dynamical scattering could thus increase the apparent spinel content from many reflection pairs. A third possible contributor to the shoulder and tail is the presence of a small number of twins of the type created by $Fm\bar{3}m$ to $R\bar{3}m$ symmetry lowering during synthesis (more details in the Supplementary Materials).

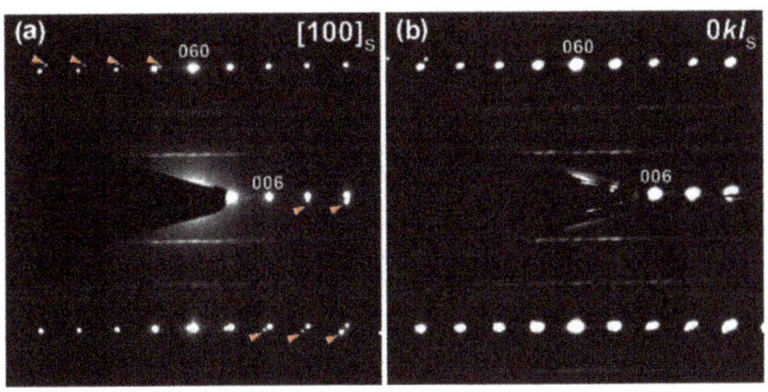

Figure 9. (**a**) $[100]_S$ in-zone PED pattern and (**b**) $0kl$ section of a 3D ED series of the same 150× cycled discharged LR-NMC crystal used for the phase ratio determination (orange arrowheads point to the reflections from another domain).

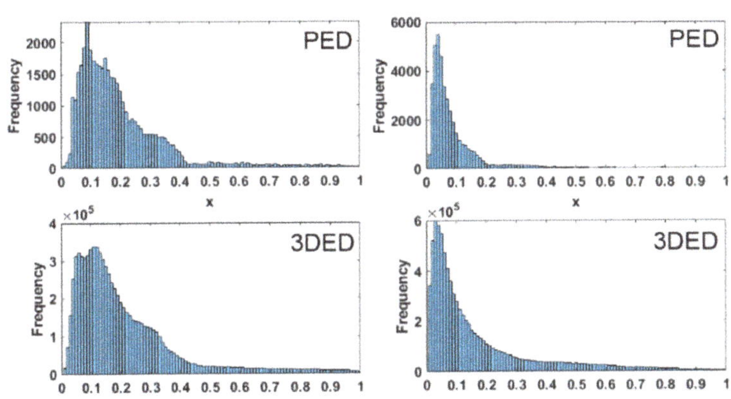

Figure 10. Histograms showing the estimated spinel content in 150× cycled discharged LR-NMC crystals. The left-side histograms show the results when including only a single honeycomb layered twin, while the right-side histograms show the improved results by including also the 120° and 240° rotation twins. The results are shown for both the PED data (**top**) and 3D ED data (**bottom**). The number of pairs contributing to each histogram bar is the frequency divided by 100.

For both the 3D ED data and the in-zone PED data, the highest peaks on the resulting histograms are best fitted using Gaussian peaks complemented by two extra smaller Gaussian peaks at higher x value describing the shoulders and tail. For both the PED in-zone data and the 3D ED data, the fit results in a peak at x = 0.04 with a half-width-half-maximum of 0.04 and a height resp. of 6418 (64 reflection pairs) and 1,069,885 (10,698 reflection pairs).

Long duration in situ synchrotron X-ray powder diffraction (SXRD) showed a continuous and irreversible increase in the number of transition metal atoms at the octahedral positions in the Li-layers–up to 5% after 100 cycles [2]. This is close to the 4% spinel content found in our study using particles that were 150 times cycled and observed in the discharged state. The slight difference might be due to the experimental errors but could also be due to transition metal atoms being at the octahedral positions in the Li-layers, but not in an ordered manner. These transition metal atoms would not contribute to spinel reflections (4% spinel from 3D ED and PED) but would contribute to the occupancy factor of those positions (5% from SXRD).

In our single-crystal electron diffraction data, the spinel reflections are present in the (*hkl*, *I*)-lists, supporting the observation from TEM imaging [1,5,25] that the spinel structure is present in the cycled LR-NMC particles. Since the conventional settings used for electron diffraction do not influence the formation of spinel, the 3D ED and PED data confirm that the spinel phase was formed during cycling and not induced by the electron beam.

The proposed method could also be applied to other combinations of phases as long as some reflections belong to only one of the two phases. For the current paper, the 3D ED data were acquired manually, but several methods are freely available that allow the taking of 3D ED series of many particles in a short time, such as fast electron diffraction tomography [38], ultra-fast electron diffraction tomography [46], fast-ADT [35], Instamatic [36,47] and InsteadMatic [37]—all part of the modern 3D ED family [14]. By combining such fast data collection with the simple phase ratio determination presented here, it should be possible to evaluate in a short time the phase ratio in a statistically relevant number of crystals or for the different stages of phase transformations taking place inside the transmission electron microscope during in situ experiments.

4. Conclusions

Electron diffraction data can be recorded without affecting the amount of spinel phase in
$Li_{1.2}Ni_{0.13}Mn_{0.54}Co_{0.13}O_2$ (LR-NMC). This allows estimation of the relative volume of different phases in LR-NMC crystallites using the intensity ratios of reflections in precession electron diffraction (PED) patterns or in three-dimensional electron diffraction (3D ED) reconstructions. There has to be a careful consideration of the reflections that will be used, to have sufficient contributions from each phase and exclude other possibly present phases, and to exclude reflection pairs for which the intensity ratio does not show monotonous behavior.

The results of the application of this phase ratio determination method on experimental data obtained from 150 times cycled LR-NMC examined in the discharged state allowed the derivation of an amount of spinel in line with the amount of transition metal atoms found in octahedral positions in the lithium layer using synchrotron X-ray powder diffraction.

Supplementary Materials: The following are available online at https://www.mdpi.com/2073-8994/13/11/1989/s1, Figure S1: overview of the main zones (and relations between them) for all phases involved, indexed in original honeycomb layered, spinel and rock salt indices and in the supercell indices. Figure S2: $[100]_S$ electron diffraction patterns calculated for, respectively, 5%, 20% and 50% of spinel, for a spherical crystal with a size of 10 × 10 × 10 supercell unit cells. Figure S3: histograms of the resulting phase ratios for the calculated (*hkl*, *I*)-lists using different sets of reflection pairs. Clarification of the occurrence of twins in the compounds and how to recognize them. Figure S4: the relationship between the parent rock salt structure and the two common layered structures. Figure S5: equivalent zones in the cubic system. Figure S6: equivalent zones in the cubic system. Table S1: Details of the two-phase refinement of the spinel content in LR-NMC from ED patterns simulated using a spherical core-shell particle.

Author Contributions: Conceptualization, M.Q., T.P. and J.H.; methodology, T.P. and J.H.; software, T.P. and M.Q.; validation, R.P., M.B., M.Q., J.H.; formal analysis, all authors; investigation, M.Q.,

R.P., M.B.; resources, J.H.; writing—original draft preparation, M.Q. and J.H.; writing—review and editing, all authors; supervision, J.H.; funding acquisition, J.H. All authors have read and agreed to the published version of the manuscript.

Funding: This research was funded by FWO, grant number G035619N and G040116N.

Institutional Review Board Statement: Not applicable.

Informed Consent Statement: Not applicable.

Data Availability Statement: The (*hkl*, *I*)-lists obtained from 3D ED and PED data are available on Zenodo at 10.5281/zenodo.5175929, accessed on 10 August 2021.

Acknowledgments: We thank Stefano Canossa, Daphne Vandemeulebroucke and Saleh Gholam for fruitful discussions.

Conflicts of Interest: The authors declare no conflict of interest. The funders had no role in the design of the study; in the collection, analyses, or interpretation of data; in the writing of the manuscript, or in the decision to publish the results.

References

1. Genevois, C.; Koga, H.; Croguennec, L.; Ménétrier, M.; Delmas, C.; Weill, F. Insight into the atomic structure of cycled lithium-rich layered oxide $Li_{1.20}Mn_{0.54}Co_{0.13}Ni_{0.13}O_2$ using HAADF STEM and electron nanodiffraction. *J. Phys. Chem. C* **2015**, *119*, 75–83. [CrossRef]
2. Kleiner, K.; Strehle, B.; Baker, A.R.; Day, S.J.; Tang, C.C.; Buchberger, I.; Chesneau, F.-F.; Gasteiger, H.A.; Piana, M. Origin of High Capacity and Poor Cycling Stability of Li-Rich Layered Oxides: A Long-Duration in Situ Synchrotron Powder Diffraction Study. *Chem. Mater.* **2018**, *30*, 3656–3667. [CrossRef]
3. Zhan, C.; Wu, T.; Lu, J.; Amine, K. Dissolution, migration, and deposition of transition metal ions in Li-ion batteries exemplified by Mn-based cathodes-A critical review. *Energy Environ. Sci.* **2018**, *11*, 243–257. [CrossRef]
4. Fan, Y.; Zhang, W.; Zhao, Y.; Guo, Z.; Cai, Q. Fundamental understanding and practical challenges of lithium-rich oxide cathode materials: Layered and disordered-rocksalt structure. *Energy Storage Mater.* **2021**, *40*, 51–71. [CrossRef]
5. Pimenta, V.; Sathiya, M.; Batuk, D.; Abakumov, A.M.; Giaume, D.; Cassaignon, S.; Larcher, D.; Tarascon, J.M. Synthesis of Li-Rich NMC: A Comprehensive Study. *Chem. Mater.* **2017**, *29*, 9923–9936. [CrossRef]
6. McCalla, E.; Abakumov, A.M.; Saubanere, M.; Foix, D.; Berg, E.J.; Rousse, G.; Doublet, M.-L.M.L.; Gonbeau, D.; Novak, P.; Van Tendeloo, G.; et al. Visualization of O-O peroxo-like dimers in high-capacity layered oxides for Li-ion batteries. *Science* **2015**, *350*, 1516–1521. [CrossRef]
7. Mauger, A.; Julien, C. Surface modifications of electrode materials for lithium-ion batteries: Status and trends. *Ionics* **2014**, *20*, 751–787. [CrossRef]
8. Lu, P.; Yan, P.; Romero, E.; Spoerke, E.D.; Zhang, J.G.; Wang, C.M. Observation of electron-beam-induced phase evolution mimicking the effect of the charge-discharge cycle in Li-rich layered cathode materials used for Li ion batteries. *Chem. Mater.* **2015**, *27*, 1375–1380. [CrossRef]
9. Lin, F.; Markus, I.M.; Nordlund, D.; Weng, T.-C.; Asta, M.D.; Xin, H.L.; Doeff, M.M. Surface reconstruction and chemical evolution of stoichiometric layered cathode materials for lithium-ion batteries. *Nat. Commun.* **2014**, *5*, 3529. [CrossRef] [PubMed]
10. Lin, F.; Markus, I.M.; Doeff, M.M.; Xin, H.L. Chemical and structural stability of lithium-ion battery electrode materials under electron beam. *Sci. Rep.* **2014**, *4*, 5694. [CrossRef] [PubMed]
11. Shukla, A.K.; Ramasse, Q.M.; Ophus, C.; Duncan, H.; Hage, F.; Chen, G. Unravelling structural ambiguities in lithium- and manganese-rich transition metal oxides. *Nat. Commun.* **2015**, *6*, 1–9. [CrossRef] [PubMed]
12. Jarvis, K.A.; Wang, C.C.; Knight, J.C.; Rabenberg, L.; Manthiram, A.; Ferreira, P.J. Formation and effect of orientation domains in layered oxide cathodes of lithium-ion batteries. *Acta Mater.* **2016**, *108*, 264–270. [CrossRef]
13. Jarvis, K.A.; Deng, Z.; Allard, L.F.; Manthiram, A.; Ferreira, P.J. Understanding structural defects in lithium-rich layered oxide cathodes. *J. Mater. Chem.* **2012**, *22*, 11550–11555. [CrossRef]
14. Gemmi, M.; Mugnaioli, E.; Gorelik, T.E.; Kolb, U.; Palatinus, L.; Boullay, P.; Hovmöller, S.; Abrahams, J.P. 3D electron diffraction: The nanocrystallography revolution. *ACS Cent. Sci.* **2019**, *5*, 1315–1329. [CrossRef] [PubMed]
15. Palatinus, L.; Corrêa, C.A.; Steciuk, G.; Jacob, D.; Roussel, P.; Boullay, P.; Klementová, M.; Gemmi, M.; Kopeček, J.; Domeneghetti, M.C.; et al. Structure refinement using precession electron diffraction tomography and dynamical diffraction: Tests on experimental data. *Acta Crystallogr. Sect. B Struct. Sci. Cryst. Eng. Mater.* **2015**, *71*, 740–751. [CrossRef] [PubMed]
16. Karakulina, O.M.; Demortière, A.; Dachraoui, W.; Abakumov, A.M.; Hadermann, J. In Situ Electron Diffraction Tomography Using a Liquid-Electrochemical Transmission Electron Microscopy Cell for Crystal Structure Determination of Cathode Materials for Li-Ion batteries. *Nano Lett.* **2018**, *18*, 6286–6291. [CrossRef]
17. Hadermann, J.; Abakumov, A.M. Structure solution and refinement of metal-ion battery cathode materials using electron diffraction tomography. *Acta Crystallogr. Sect. B Struct. Sci. Cryst. Eng. Mater.* **2019**, *75*, 485–494. [CrossRef] [PubMed]

18. Hadermann, J.; Abakumov, A.; Van Rompaey, S.; Perkisas, T.; Filinchuk, Y.; Van Tendeloo, G. Crystal structure of a lightweight borohydride from submicrometer crystallites by precession electron diffraction. *Chem. Mater.* **2012**, *24*, 3401–3405. [CrossRef]
19. Orlova, E.D.; Savina, A.A.; Abakumov, S.A.; Morozov, A.V.; Abakumov, A.M. Comprehensive Study of Li+/Ni2+ Disorder in Ni-Rich NMCs Cathodes for Li-Ion Batteries. *Symmetry* **2021**, *13*, 1628. [CrossRef]
20. Zhao, H.; Krysiak, Y.; Hoffmann, K.; Barton, B.; Molina-Luna, L.; Neder, R.B.; Kleebe, H.-J.; Gesing, T.M.; Schneider, H.; Fischer, R.X.; et al. Elucidating structural order and disorder phenomena in mullite-type $Al_4B_2O_9$ by automated electron diffraction tomography. *J. Solid State Chem.* **2017**, *249*, 114–123. [CrossRef]
21. Krysiak, Y.; Barton, B.; Marler, B.; Neder, R.B.; Kolb, U. *Ab initio* structure determination and quantitative disorder analysis on nanoparticles by electron diffraction tomography. *Acta Crystallogr. Sect. A Found. Adv.* **2018**, *74*, 93–101. [CrossRef] [PubMed]
22. Brázda, P.; Palatinus, L.; Klementová, M.; Buršík, J.; Knížek, K. Mapping of reciprocal space of $La_{0.30}CoO_2$ in 3D: Analysis of superstructure diffractions and intergrowths with Co_3O_4. *J. Solid State Chem.* **2015**, *227*, 30–34. [CrossRef]
23. Lanza, A.; Margheritis, E.; Mugnaioli, E.; Cappello, V.; Garau, G.; Gemmi, M. Nanobeam precession-assisted 3D electron diffraction reveals a new polymorph of hen egg-white lysozyme. *IUCrJ* **2019**, *6*, 178–188. [CrossRef]
24. Paulus, A.; Hendrickx, M.; Bercx, M.; Karakulina, O.M.; Kirsanova, M.A.; Lamoen, D.; Hadermann, J.; Abakumov, A.M.; Van Bael, M.K.; Hardy, A. An in-depth study of Sn substitution in Li-rich/Mn-rich NMC as a cathode material for Li-ion batteries. *Dalt. Trans.* **2020**, *49*, 10486–10497. [CrossRef]
25. Proffen, T.; Neder, R.B. DISCUS: A program for diffuse scattering and defect-structure simulation. *J. Appl. Crystallogr.* **1997**, *30*, 171–175. [CrossRef]
26. Neder, R.B.; Proffen, T. *Diffuse Scattering and Defect Structure Simulations: A Cook Book Using the Program DISCUS*; Oxford University Press: Oxford, UK, 2008; Volume 11, ISBN 9780191715563.
27. Palatinus, L. PETS—Program for Analysis of Electron Diffraction Data. Available online: http://pets.fzu.cz/ (accessed on 1 August 2021).
28. Palatinus, L.; Brázda, P.; Jelínek, M.; Hrdá, J.; Steciuk, G.; Klementová, M. Specifics of the data processing of precession electron diffraction tomography data and their implementation in the program PETS2.0. *Acta Crystallogr. Sect. B Struct. Sci. Cryst. Eng. Mater.* **2019**, *75*, 512–522. [CrossRef]
29. Petříček, V.; Dušek, M.; Palatinus, L. Crystallographic Computing System JANA2006: General features. *Z. Krist. Cryst. Mater.* **2014**, *229*, 345–352. [CrossRef]
30. Wojdyr, M. Fityk: A general-purpose peak fitting program. *J. Appl. Crystallogr.* **2010**, *43*, 1126–1128. [CrossRef]
31. Karakulina, O.M.; Khasanova, N.R.; Drozhzhin, O.A.; Tsirlin, A.A.; Hadermann, J.; Antipov, E.V.; Abakumov, A.M. Antisite Disorder and Bond Valence Compensation in Li2FePO4F Cathode for Li-Ion Batteries. *Chem. Mater.* **2016**, *28*, 7578–7581. [CrossRef]
32. Drozhzhin, O.A.; Sumanov, V.D.; Karakulina, O.M.; Abakumov, A.M.; Hadermann, J.; Baranov, A.N.; Stevenson, K.J.; Antipov, E.V. Switching between solid solution and two-phase regimes in the Li1-xFe1-yMnyPO4 cathode materials during lithium (de)insertion: Combined PITT, in situ XRPD and electron diffraction tomography study. *Electrochim. Acta* **2016**, *191*, 149–157. [CrossRef]
33. Mikhailova, D.; Karakulina, O.M.; Batuk, D.; Hadermann, J.; Abakumov, A.M.; Herklotz, M.; Tsirlin, A.A.; Oswald, S.; Giebeler, L.; Schmidt, M.; et al. Layered-to-Tunnel Structure Transformation and Oxygen Redox Chemistry in $LiRhO_2$ upon Li Extraction and Insertion. *Inorg. Chem.* **2016**, *55*, 7079–7089. [CrossRef]
34. Fedotov, S.S.; Khasanova, N.R.; Samarin, A.S.; Drozhzhin, O.A.; Batuk, D.; Karakulina, O.M.; Hadermann, J.; Abakumov, A.M.; Antipov, E.V. $AVPO_4F$ (A = Li, K): A 4 V Cathode Material for High-Power Rechargeable Batteries. *Chem. Mater.* **2016**, *28*, 411–415. [CrossRef]
35. Plana-Ruiz, S.; Krysiak, Y.; Portillo, J.; Alig, E.; Estradé, S.; Peiró, F.; Kolb, U. Fast-ADT: A fast and automated electron diffraction tomography setup for structure determination and refinement. *Ultramicroscopy* **2020**, *211*, 112951. [CrossRef]
36. Cichocka, M.O.; Ångström, J.; Wang, B.; Zou, X.; Smeets, S. High-throughput continuous rotation electron diffraction data acquisition via software automation. *J. Appl. Crystallogr.* **2018**, *51*, 1652–1661. [CrossRef] [PubMed]
37. Wang, B.; Zou, X.; Smeets, S. Automated serial rotation electron diffraction combined with cluster analysis: An efficient multi-crystal workflow for structure determination. *IUCrJ* **2019**, *6*. [CrossRef] [PubMed]
38. Gemmi, M.; La Placa, M.G.I.; Galanis, A.S.; Rauch, E.F.; Nicolopoulos, S. Fast electron diffraction tomography. *J. Appl. Crystallogr.* **2015**, *48*, 718–727. [CrossRef]
39. Gruene, T.; Wennmacher, J.T.C.; Zaubitzer, C.; Holstein, J.J.; Heidler, J.; Fecteau-Lefebvre, A.; De Carlo, S.; Müller, E.; Goldie, K.N.; Regeni, I.; et al. Rapid Structure Determination of Microcrystalline Molecular Compounds Using Electron Diffraction. *Angew. Chemie—Int. Ed.* **2018**, *57*, 16313–16317. [CrossRef]
40. Palatinus, L.; Brázda, P.; Boullay, P.; Perez, O.; Klementová, M.; Petit, S.; Eigner, V.; Zaarour, M.; Mintova, S. Hydrogen positions in single nanocrystals revealed by electron diffraction. *Science* **2017**, *355*, 166–169. [CrossRef]
41. Kolb, U.; Gorelik, T.E.; Mugnaioli, E.; Stewart, A. Structural Characterization of Organics Using Manual and Automated Electron Diffraction. *Polym. Rev.* **2010**, *50*, 385–409. [CrossRef]
42. Kodjikian, S.; Klein, H. Low-dose electron diffraction tomography (LD-EDT). *Ultramicroscopy* **2019**, *200*, 12–19. [CrossRef] [PubMed]

43. Steciuk, G.; David, A.; Petříček, V.; Palatinus, L.; Mercey, B.; Prellier, W.; Pautrat, A.; Boullay, P. Precession electron diffraction tomography on twinned crystals: Application to $CaTiO_3$ thin films. *J. Appl. Crystallogr.* **2019**, *52*, 626–636. [CrossRef]
44. Mohanty, D.; Huq, A.; Payzant, E.A.; Sefat, A.S.; Li, J.; Abraham, D.P.; Wood, D.L.; Daniel, C. Neutron Diffraction and Magnetic Susceptibility Studies on a High-Voltage $Li_{1.2}Mn_{0.55}Ni_{0.15}Co_{0.10}O_2$ Lithium Ion Battery Cathode: Insight into the Crystal Structure. *Chem. Mater* **2013**, *25*. [CrossRef]
45. Mosbah, A.; Verbaere, A.; Tournoux, M. Phases LixMnO2λ rattachees au type spinelle. *Mater. Res. Bull.* **1983**, *18*, 1375–1381. [CrossRef]
46. Simancas, J.; Simancas, R.; Bereciartua, P.J.; Jorda, J.L.; Rey, F.; Corma, A.; Nicolopoulos, S.; Pratim Das, P.; Gemmi, M.; Mugnaioli, E. Ultrafast Electron Diffraction Tomography for Structure Determination of the New Zeolite ITQ-58. *J. Am. Chem. Soc.* **2016**, *138*, 10116–10119. [CrossRef]
47. Smeets, S.; Wang, B.; Cichocka, M.O.; Ångström, J.; Wan, W. Instamatic. Available online: http://doi.org/10.5281/zenodo.1090389 (accessed on 1 August 2021).

symmetry

MDPI

Article

Electron Diffraction Study of the Space Group Variation in the Al–Mn–Pt T-Phase

Rimon Tamari [1], Benjamin Grushko [2,3] and Louisa Meshi [1,*]

1 Department of Materials Engineering, Ben Gurion University of the Negev, Beer Sheva 8410501, Israel; rimonta@post.bgu.ac.il
2 MaTecK GmbH, 52428 Julich, Germany; B.Grushko@fz-juelich.de
3 Peter-Grünberg-Institut, Forschungszentrum Jülich, 52425 Julich, Germany
* Correspondence: louisa@bgu.ac.il; Tel.: +972-8-647-2576

Abstract: Binary high temperature "Al_3Mn" (T-phase) and its extensions in ternary systems were the subjects of numerous crystallographic investigations. The results were ambiguous regarding the existence or lack of the center of symmetry: both $Pna2_1$ and $Pnam$ space groups were reported. Our research on the Al–Mn–Pt T-phase allowed concluding that inside a continuous homogeneity region, the structure of the Al-rich T-phase (e.g., $Al_{78}Mn_{17.5}Pt_{4.5}$) belongs to the non-centrosymmetric $Pna2_1$ space group, while the structure of the Al-poor T-phase (such as $Al_{71.3}Mn_{25.1}Pt_{3.6}$) is centrosymmetric, i.e., $Pnam$. Following metallurgical and crystallographic considerations, the change in the symmetry was explained.

Keywords: Al_3Mn; space group; Convergent Beam Electron Diffraction (CBED); center of symmetry; aluminide

1. Introduction

The Al–Mn so-called T-phase, also known as high-temperature "Al_3Mn", is formed in a compositional region between ~25 and 29 at. % Mn and a temperature range of 895–1002 °C (see [1] and references therein). The compositional region of the T-phase widely extends in numerous ternary Al–MnM alloy systems, where M is transition metal (see [2] and references therein), which results in a significant decrease in its stability temperature. Furthermore, it was argued that the binary stoichiometry of the T-phase is $Al_{124}Mn_{32}$, which is outside of its homogeneity region [2]. Reports on the atomic structure of the T-phase are puzzling. The history of the structure determination of the T-phase is a perfect example of a dispute about the accuracy of the estimation of the space group and corresponding atomic models. In 1938, the geometry of the T-phase's unit cell was reported by Hoffmann as orthorhombic [3]. In 1961, the centrosymmetric $Pnam$ space group was chosen to describe its symmetry [4], while in 1992, in a series of publications by Li et al. (see [5] and references therein), the non-centrosymmetric $Pna2_1$ was argued. In 1993, its atomic model was proposed again in the framework of the $Pnam$ space group [6], while in 1994, Shi et al. [7] insisted on $Pna2_1$. In 1995, Pavlyuk et al. [8] succeeded in refining T phase's structure in the framework of the $Pnam$ space group but failed when using $Pna2_1$, which, according to these authors, would not give significantly better results. The binary $Pnam$ prototype was chosen as a start for the structure refinement of the ternary T-phases [9–13]. Despite the common starting model, their final atomic models varied. The major discrepancies were found in the occupancies and positions of transition metals.

Recently, the extension of the T-phase region in the Al–Mn–Pt alloy system up to ~$Al_{78}Mn_{17.5}Pt_{4.5}$ and its stability at ternary compositions down to at least 800 °C were reported in [2]. A further study confirmed its stability also at 700 °C (to be published). The following lattice parameters were determined for the above-mentioned composition: a = 14.720(4) Å, b = 12.628(2) Å, c = 12.545(3) Å [2].

By using electron diffraction methods, a partial atomic model of the $Al_{78}Mn_{17.5}Pt_{4.5}$ T-phase was proposed in the framework of the non-centrosymmetric $Pna2_1$ space group [2]. The current research was undertaken with the purpose of relating the above-mentioned inconsistency in the space group to differences in the composition of the ternary T-phase. It should be mentioned that ternary extensions of the T-phase, discussed in [2], are much wider than the above-mentioned binary T-phase region. We performed a study of the symmetry of the Al–Mn–Pt T-phase varying the annealing temperature (700 vs. 800 °C) as well as its composition: $Al_{78}Mn_{17.5}Pt_{4.5}$ vs. $Al_{71.3}Mn_{25.1}Pt_{3.6}$. It was found that, regardless of the annealing temperature, the $Al_{78}Mn_{17.5}Pt_{4.5}$ T-phase is non-centrosymmetric, while the unit cell of the $Al_{71.3}Mn_{25.1}Pt_{3.6}$ T-phase can be described by the centrosymmetric *Pnam* space group. Furthermore, a direct structural relationship between the two options (centrosymmetric vs. non-centrosymmetric) is proposed.

2. Materials and Methods

Two Al–Mn–Pt compositions (see Table 1) were selected from the opposite Al limits of the compositional region of the T-phase. The alloys were produced by levitation induction melting in a water-cooled copper crucible under a pure Ar atmosphere. The purity of Al was 99.999%, Mn 99.99%, and Pt 99.9%. For homogeneity, the samples were annealed under a vacuum of 9×10^{-7} mBar.

Table 1. Al–Mn–Pt alloys studied in this research.

Nominal Composition	Heat Treatment Details
$Al_{78}Mn_{17.5}Pt_{4.5}$	800 °C for 622 h
$Al_{78}Mn_{17.5}Pt_{4.5}$	700 °C for 984 h
$Al_{71.3}Mn_{25.1}Pt_{3.6}$	800 °C for 649 h

The alloys were ground into powder using an agate mortar and pestle, dispersed in isopropanol, and stirred in the ultrasonic bath. Each suspension was drop cast on a carbon-coated Cu grid. The samples were studied using the JEOL JEM-2100 Transmission Electron Microscope (TEM) operating at 200 kV. A convergence angle of about 2×10^{-2} rad was used to acquire Convergent Beam Electron Diffraction (CBED) patterns. A condenser aperture of 20 μm in diameter was applied to eliminate the overlapping of the CBED disks.

The *Pnam* and $Pna2_1$ space groups, discussed here, exhibit the same extinction conditions: k + l = 2n for the {0kl} type reflections; h = 2n for the {h0l} and {h00} type reflections; k = 2n and l = 2n for the {0k0} and {00l} type reflections [14]. These space groups could not be distinguished by powder X-ray diffraction and conventional selected area electron diffraction (SAED). On the other hand, the CBED method has the power to perform this, differentiating between not only the extinction conditions but also the point groups. It is based on an evaluation of the highly sensitive to symmetry Zero Order Laue Zone (ZOLZ) and the whole patterns (WP), containing the Higher-Order Laue Zones (HOLZ) in addition to ZOLZ. These patterns can be 3D or 2D, depending on the information seen or not seen in the CBED disks. In some cases, even the 2D information found in the CBED patterns is sufficient for an estimation of the correct symmetry. The methodology, along with the definition of the types of patterns used, can be found in [15].

The *Pnam* and $Pna2_1$ space groups belong to the mmm and *mm*2 point groups, respectively, and as stated, exhibit different symmetries of the CBED patterns taken along the high symmetry axes [15]. During the capturing of the whole patterns, due to the above-mentioned large unit cell parameters of the T-phase, it was impossible to avoid the overlapping of the CBED disks. Thus, microdiffraction patterns with almost parallel beams were taken in the nano-beam mode. For convenience, Table 2 is reproduced from [15], summarizing the symmetry of the CBED patterns expected at the high symmetry axes for the *mmm* and *mm*2 point groups.

Table 2. Two-dimensional and 3D bright field (BF) and whole pattern (WP) symmetries of the *mmm* and *mm*2 point groups. This table was reproduced from [15].

	[100]	[10]	[1]	[u0w]	[uv0]	[0vw]
mmm	2D BF (2*mm*) 2D WP (2*mm*) 3D BF 2*mm* 3D WP 2*mm*	2D BF (2*mm*) 2D WP (2*mm*) 3D BF 2*mm* 3D WP 2*mm*	2D BF (2*mm*) 2D WP (2*mm*) 3D BF 2*mm* 3D WP 2*mm*	2D BF (2*mm*) 2D WP (2*mm*) 3D BF *m* 3D WP *m*	2D BF (2*mm*) 2D WP (2*mm*) 3D BF *m* 3D WP *m*	2D BF (2*mm*) 2D WP (2*mm*) 3D BF *m* 3D WP *m*
*mm*2	2D BF (2*mm*) 2D WP (*m*) 3D BF 2*mm* 3D WP *m*	2D BF (2*mm*) 2D WP (*m*) 3D BF 2*mm* 3D WP *m*	2D BF (2*mm*) 2D WP (2*mm*) 3D BF 2*mm* 3D WP 2*mm*	2D BF (2*mm*) 2D WP (*m*) 3D BF *m* 3D WP *m*	2D BF (2*mm*) 2D WP (*m*) 3D BF *m* 3D WP 1	2D BF (2*mm*) 2D WP (*m*) 3D BF *m* 3D WP *m*

3. Results and Discussion

Primarily, the effect of the structure relaxation on the symmetry of the unit cell was studied. It was previously reported [2] that an $Al_{78}Mn_{17.5}Pt_{4.5}$ alloy, annealed at 800 °C, contained inside the T-phase's matrix fine precipitation of a second phase, formed during cooling from the annealing temperature. This second phase was identified as isostructural to the so-called R-phase, reported to coexist with the T-phase in the Al–Mn–Pd alloy system, where $Al_{78.6}Mn_{15.6}Pd_{5.7}$ R-phase was found to have the *Cmcm* space group and following lattice parameters a = 7.76 Å, b = 23.88 Å, c = 12.43 Å [9,16]. The presence of this phase indicates that the precipitates-free regions of the T-phase, which were used for the structure determination in [2], could be supersaturated. Thus, the structure in these regions could have a different space group than that ascribed to the binary T-phase due to the corresponding internal strain. Therefore, the same alloy was additionally annealed at 700 °C to complete the precipitation of the R-phase, which would result in a relaxation of the T-phase's matrix. By following this annealing, the resultant material contained equiaxed grains of both phases, exceeding 100 nm in diameter, which is sufficient for electron diffraction experiments.

The unit cell parameters of the T- and R-phases are closely related [5,9,12] and, hence, many interplanar distances (d_{hkl}) overlap. Therefore, at many orientations, electron diffraction patterns could be indexed in terms of both phases. The most effective way to distinguish between them is either by comparing their strong reflections distribution at the principal axes [16] or by the presence of the reflections with the interplanar distance of ~9.5 Å ({110} of the T-phase), which do not appear in the R-phase interplanar spacing list. Therefore, the CBED analysis was performed using the $[1\bar{1}0]$ zone axis patterns of the T-phase. According to Table 2, the difference between the discussed point groups (*mmm* vs. *mm*2) along the $[1\bar{1}0]$ orientation can be observed using the 2D WP patterns. This means that only intensity distribution of the CBED disks is considered while assessing the symmetry, i.e., without taking features as excess and deficiency lines, the ZOLZ fringes, etc., into account. This information is available on ZOLZ CBED patterns displaying only ZOLZ disks. The mmm point group will impose the 2*mm* symmetry on the ZOLZ CBED pattern, while the *m* symmetry of this pattern will indicate that the *mm*2 point group is the correct one.

The 2D ZOLZ patterns of the $Al_{78}Mn_{17.5}Pt_{4.5}$ T-phase annealed at 700 and 800 °C are shown in Figure 1a,b, respectively. Regardless of the annealing temperature, i.e., in either supersaturated or relaxed state, the symmetry of the T-phase remained the same. In both cases, symmetry was evaluated as *m*, indicating the *mm*2 point group, and subsequently the $Pna2_1$ space group, as reported in [2].

The change in the symmetry as a function of composition can be seen clearly in Figure 1c. For the $Al_{71.3}Mn_{25.1}Pt_{3.6}$ composition (i.e., with a visibly lower Al/Mn ratio), the symmetry of the ZOLZ pattern is 2*mm*, indicating the *mmm* point group and, subsequently, the *Pnam* space group, which is in line with reported for the binary T-"Al_3Mn" phase [6].

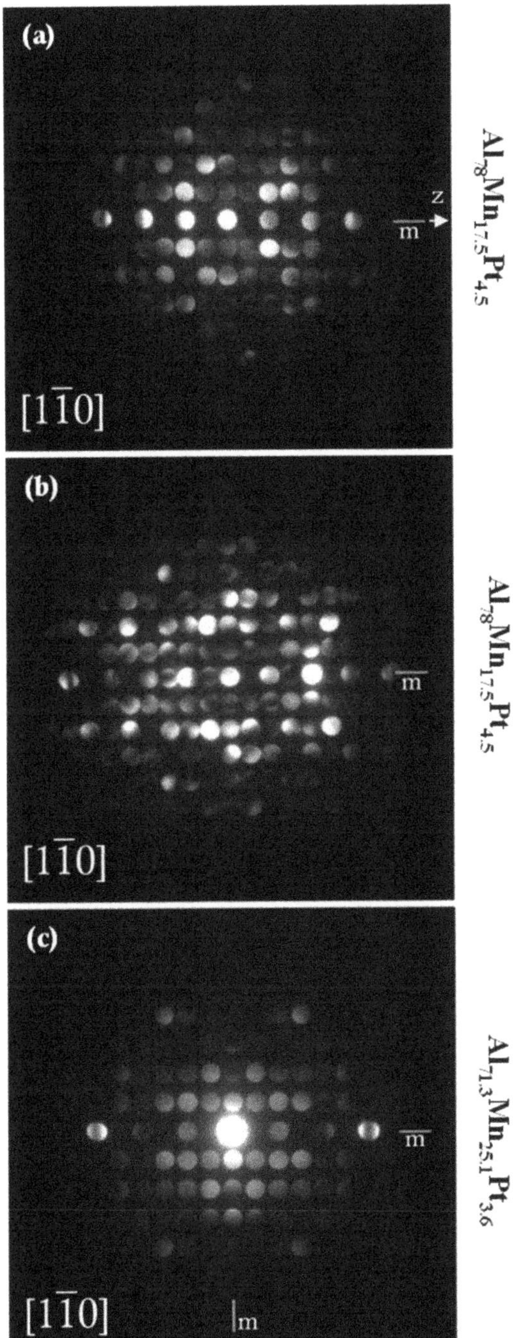

Figure 1. Two-dimensional [1$\bar{1}$0] ZOLZ CBED patterns taken from the $Al_{78}Mn_{17.5}Pt_{4.5}$ T-phase, annealed at 800 °C (**a**) and 700 °C (**b**) and from the $Al_{71.3}Mn_{25.1}Pt_{3.6}$ T-phase, annealed at 800 °C (**c**). Mirrors are denoted by *m*. [001] direction of the reciprocal space is marked as *z* on the upper image. All images were aligned to each other.

The structure of the T-phase with the *Pnam* space group can be described as a stacking of flat and puckered layers [5]. If these flat layers are not strictly flat, their mirror symmetry is destroyed, resulting in the $Pna2_1$ space group, and a displacement of even ±0.005 from z = 1/4 is sufficient for this. Here, (x, y, 1/4) is the 4*c* special Wyckoff position in *a-cb* setting of the *Pnma* space group, i.e., *Pnam*, while in the $Pna2_1$ space group, this position is general, of the (x, y, z) type. In a wide compositional region, the replacement between several larger Al and smaller Mn atoms could easily result in such displacements, which would be more reliably recognized at compositions with an extremal Al/Mn ratio. Obviously, this is also the question of the experimental resolution.

On the other hand, due to the change in the space group, some splitting of the corresponding general 8*d* Wyckoff positions would be inevitable since, in the $Pna2_1$ space group, only the 4a positions exist. Since the maximal translationengleiche subgroup of *Pnam* is $Pna2_1$ [17], the transition from one to another will cause the described above splitting of each general 8*d* Wyckoff position into two general 4*a* positions: one with the same coordinates, turning to be of the (x, y, z) type, and the other (0.5 − x, 0.5 + y, 1 − z)). During this transition, each special 4*c* position (of the (x, y, 1/4) type in the *Pnam* space group) turns into a general position (x, y, z) of the 4*a* type in the $Pna2_1$ space group [14,17].

Considering the stoichiometry of the T-phase as binary $Al_{124}Mn_{32}$, this is plausible to place the 8 Pt atoms in $Al_{78}Mn_{17.5}Pt_{4.5}$ at specific Mn cites (see [2]), while by decreasing Al concentration, some Al sites would be (partially) occupied by Mn, as it is in the case of the current structural models of the T-phase [6,8]. Our results are in favor of a suggestion that the increase in symmetry is rather caused by the replacement of Al by Mn, while the Mn/Pt substitution seems to have no influence on the symmetry.

4. Conclusions

The current research focuses on the unambiguous determination of the space group of the Al–Mn–Pt T-phase. Our work was motivated by the discussion on the symmetry of the T-phase, reported in binary Al–Mn and various ternary Al–Mn–M (where M is transition metal) alloy systems, which was held for several decades. By using the CBED method, it was found that the increase in the Al/Mn ratio leads to the change from the *Pnam* to $Pna2_1$ space group. On the other hand, regardless of the annealing temperature, the symmetry of the T-phase remained unchanged. Following metallurgical and crystallographic considerations, an explanation for the symmetry change (centrosymmetric vs. non-centrosymmetric) was proposed.

Author Contributions: Conceptualization, L.M.; methodology, L.M.; software, R.T.; validation, R.T., B.G. and L.M.; formal analysis, R.T.; investigation, R.T.; resources, L.M.; data curation, L.M. and B.G.; writing—original draft preparation, R.T.; writing—review and editing, L.M. and B.G.; visualization, R.T.; supervision, L.M.; funding acquisition, L.M. All authors have read and agreed to the published version of the manuscript.

Funding: This research received no external funding.

Institutional Review Board Statement: Not applicable.

Informed Consent Statement: Not applicable.

Data Availability Statement: Not applicable.

Conflicts of Interest: The authors declare no conflict of interest.

References

1. McAlister, A.J.; Murray, J.L. The Aluminum-Manganese system. *Bull. Alloy. Phase Diagr.* **1987**, *8*, 438–447. [CrossRef]
2. Tamari, R.; Grushko, B.; Meshi, L. Structural study of $Al_{78}Mn_{17.5}Pt_{4.5}$ and (re) constitution of the Al–Mn–Pt system in its vicinity. *J. Alloys Compd.* **2021**, *861*, 158328. [CrossRef]
3. Hoffmann, W. Roentgenographic methods in the investigation of Alumium alloys. *Aluminium* **1938**, *20*, 865–872.
4. Taylor, M.A. The space group of $MnAl_3$. *Acta Crystallogr.* **1961**, *14*, 84. [CrossRef]

5. Li, X.Z.; Shi, D.; Kuo, K.H. Structure of Al_3Mn, an orthorhombic approximant of the decagonal quasicrystal. *Philos. Mag. Part B* **1992**, *66*, 331–340. [CrossRef]
6. Hiraga, K.; Kaneko, M.; Matsuo, Y.; Hashimoto, S. The structure of Al_3Mn: Close relationship to decagonal quasicrystals. *Philos. Mag. Part B* **1993**, *67*, 193–205. [CrossRef]
7. Shi, N.C.; Li, X.Z.; Ma, Z.S.; Kuo, K.H. Crystalline phases related to a decagonal quasicrystal. I. A single-crystal X-ray diffraction study of the orthorhombic Al_3Mn phase. *Acta Crystallogr. B* **1994**, *50*, 22–30. [CrossRef]
8. Pavlyuk, V.V.; Yanson, T.I.; Bodak, O.I.; Černý, R.; Gladyshevskii, R.E.; Yvon, K.; Stepien-Damm, J. Structure refinement of orthorhombic $MnAl_3$. *Acta Crystallogr. C* **1995**, *51*, 792–794. [CrossRef]
9. Audier, M.; Durand-Charre, M.; de Boissieu, M. AlPdMn phase diagram in the region of quasicrystalline phases. *Philos. Mag. Part B* **1993**, *68*, 607–618. [CrossRef]
10. Matsuo, Y.; Yamamoto, K.; Iko, Y. Structure of a new orthorhombic crystalline phase in the Al-Cr-Pd alloy system. *Philos. Mag. Lett.* **1997**, *75*, 137–142. [CrossRef]
11. Matsuo, Y.; Kaneko, M.; Yamanoi, T.; Kaji, N.; Sugiyama, K.; Hiraga, K. The structure of an Al_3Mn-type Al_3(Mn, Pd) crystal studied by single-crystal X-ray diffraction analysis. *Philos. Mag. Lett.* **1997**, *76*, 357–362. [CrossRef]
12. Klein, H.; Boudard, M.; Audier, M.; de Boissieu, M.; Vincent, H.; Beraha, L.; Duneau, M. The T-Al_3(Mn, Pd) quasicrystalline approximant: Chemical order and phason defects. *Philos. Mag. Lett.* **1997**, *75*, 197–208. [CrossRef]
13. Furihata, J.I.; Okabe, T. Crystalline approximants of decagonal quasicrystals in the Al-Cu-Cr system. *Microscopy* **1999**, *48*, 761–766. [CrossRef]
14. Hahn, T. (Ed.) *International Tables for Crystallography. Volume A: Space-Group Symmetry*; Springer: Berlin, Germany, 2005; p. 911.
15. Morniroli, J.P. Atlas of Electron Diffraction Zone Axis Patterns. 2013. Available online: http://www.electron-diffraction.fr/ (accessed on 14 November 2021).
16. Balanetskyy, S.; Meisterernst, G.; Heggen, M.; Feuerbacher, M. Reinvestigation of the Al–Mn–Pd alloy system in the vicinity of the T- and R-phases. *Intermetallics* **2008**, *16*, 71–87. [CrossRef]
17. Wondratschek, H.; Muller, U. (Eds.) *International Tables of Crystallography, Volume A1: Symmetry Relations Between Space Groups*; Kluwer Academic Publishers: Amsterdam, The Netherlands, 2004; p. 731.

symmetry

MDPI

Article

Low-Dose Electron Crystallography: Structure Solution and Refinement

Holger Klein *, Stéphanie Kodjikian, Emre Yörük and Pierre Bordet

Institut Néel, Université Grenoble Alpes, CNRS, 38000 Grenoble, France; stephanie.kodjikian@neel.cnrs.fr (S.K.); emre.yoruk@neel.cnrs.fr (E.Y.); Pierre.Bordet@neel.cnrs.fr (P.B.)

* Correspondence: holger.klein@neel.cnrs.fr

Abstract: There is a wealth of materials that are beam sensitive and only exist in nanometric crystals, because the growth of bigger crystals is either impossible or so complicated that it is not reasonable to spend enough time and resources to grow big crystals before knowing their potential for research or applications. This difficulty is encountered in minerals, zeolites, metal-organic frameworks or molecular crystals, including pharmaceuticals and biological crystals. In order to study these crystals a structure determination method for beam sensitive crystals of nanometric size is needed. The nanometric size makes them destined for electron diffraction, since electrons interact much more strongly with matter than X-rays or neutrons. In addition, for the same amount of beam damage, electron diffraction yields more information than X-rays. The recently developed low-dose electron diffraction tomography (LD-EDT) not only combines the advantages inherent in electron diffraction, but is also optimized for minimizing the electron dose used for the data collection. The data quality is high, allowing not only the solution of complex unknown structures, but also their refinement taking into account the dynamical diffraction effects. Here we present several examples of crystals solved and refined by this method. The range of the crystals presented includes two synthetic oxides, $Sr_5CuGe_9O_{24}$ and $(Na_{2/3}Mn_{1/3})_3Ge_5O_{12}$, a natural mineral (bulachite), and a metal organic framework (Mn-formiate). The dynamical refinement can be successfully performed on data sets that needed less than 0.1 $e^-/\text{Å}^2$ for the entire data set.

Keywords: electron crystallography; beam sensitive materials; structure solution; structure refinement; 3D electron diffraction

1. Introduction

The physical, chemical, and biological properties and functionalities of a material can depend very strongly on the crystallographic structure of the compound and not only on its chemical composition. In material sciences the physical properties of graphite and diamond are fundamentally different while both are constituted of only carbon atoms. The 3-dimensional bonding scheme and electronic structure of diamond lead to a transparent crystal of high thermal and low electrical conductivity while the 2-dimensional bonding in graphite is responsible for the high light absorbance, and low thermal and relatively high electric conductivity. In chemistry and life sciences enantiomers, molecules with the same chemical formula, but a different "handedness" behave differently as for example D-asparagine has a sweet taste, while the natural occurring L-asparagine is tasteless. In addition, for pharmaceuticals the exact crystalline state of an active agent is often critical for its bio-availability.

It is therefore paramount to gain as much insight into the structure of a compound as possible in order to fully understand its properties. Over the last century, many different X-ray diffraction (and later on neutron diffraction) techniques have been developed for this purpose and an astounding number of crystal structures have been solved. More recently, electron diffraction in a transmission electron microscope (TEM) has been employed in crystallography, especially in cases where it was impossible to grow single crystals big enough for X-ray diffraction. In these cases, one can take advantage of the strong interaction of electrons with matter which allows using crystals that are only a few tens of nanometers in diameter. Especially for beam sensitive materials another advantage of electron over X-ray diffraction is the fact that for the same amount of beam damage, electron diffraction yields more information than X-rays [1].

Especially 3D electron diffraction methods have been very successful in the recent years. Different variants of 3D electron diffraction (3D ED) have been developed using either stepwise acquisition of diffraction frames [2–4] or data acquisition while continuously rotating the sample (continuous rotation electron diffraction) [5]. In both cases the acquisition is automatized and therefore more rapid than manually recording the data, however, the displacement of the sample during the tilting meant that a re-centering step had to be included in between recorded frames. The so-called fast electron diffraction tomography [6] overcame this constraint by calibrating the displacement of the sample during the tilt before the experiment and then programming the incident beam shift accordingly, so that the beam followed the displacement of the crystal. A comprehensive review of 3D ED method can be found in a paper by Gemmi et al. [7].

All of these methods have allowed structure solutions of different types of crystals including beam sensitive materials like small molecules or zeolites and also the refinement of these structures [8,9]. However, there are structures that are even more beam sensitive such as some metal-organic-frameworks (MOF) or biological crystals. Even in these cases some structures have been solved from 3D ED [10,11]. Beam damage results in a loss of crystallinity and lower resolution data as reflections lose their intensity [12]. For these materials, a 3D ED method where the irradiation dose is minimized is necessary. Low-dose electron diffraction tomography (LD-EDT) [13] is such a method and in this manuscript we present a variety of examples of crystal structures that have been solved with this method. While the structure solution does not require very precisely measured diffraction intensities [14], the quality of the data collection is important for the subsequent structure refinement. We also present the refinement results using the LD-EDT data.

2. Experimental

For the data acquisition the samples were dispersed in powder form on a copper grid covered with a holey amorphous carbon film. The electron diffraction data were obtained on a Philips CM300ST transmission electron microscope (TEM). The TEM is equipped with a Bruker AXS EDX, a Nanomegas spinning star precession module, a TVIPS F416 CMOS camera, and an in-house modified single tilt sample holder that allows tilt angles up to $\pm 55°$.

The chosen LD-EDT method [13] is based on a tomography of the reciprocal space. An isolated particle of the powder is selected for data acquisition. The size of the crystal should typically be between several tens of nanometers and around a micron. The crystal is then carefully adjusted at the eucentric height of the goniometer in order to move as little as possible during the tomography. Without any prior orientation of the crystal, it is then placed inside a large selected area aperture (typically 2 μm in apparent diameter). The tomography data are then recorded over a large angular domain (usually between $60°$ and $110°$) in discrete steps of typically $1°$. In order to fully integrate the intensities over the reciprocal rods we applied precession to the diffraction [15] with a precession half-angle of $1.05°$. In order to minimize the electron dose, the sample undergoes during the data acquisition, the electron beam was blanked at all times except for the recording of the diffraction frames. For more details about the LD-EDT method see [13].

After each tomography an image of the vacuum close to the sample was recorded in the same illumination conditions as used for a single frame of the tomography. Knowing the number of counts on the camera created by a single incident electron the electron dose for an individual diffraction frame was determined. The total dose used for the experiment is then obtained by multiplying the dose per frame with the number of frames.

The values of the experimental parameters are given in Table 1. Starting from the recorded frames the indexing and reconstruction of the 3-dimensional reciprocal lattice was done using the PETS2 software [16]. The structure solution was then attempted by SIR2014 [17] and Superflip [18] embedded into Jana2006 [19]. For the refinement based on the dynamical theory of diffraction Jana2006 was used.

Table 1. Values of the experimental parameters used for the data collection in LD-EDT.

Crystal	Tilt Step	Precession Angle	Number of Frames	Exposure Time per Frame	Total Dose
$Sr_5CuGe_9O_{24}$	1	1.05	60	100 ms	$0.1\ e^-/Å^2$
Mn-formiate $[Mn(HCOO)_2(H_2O)_2]_\infty$	1	1.05	60	200 ms	$0.15\ e^-/Å^2$
Garnet $(Na_{2/3}Mn_{1/3})_3Ge_5O_{12}$	1	1.05	95	260 ms	$0.13\ e^-/Å^2$
Bulachite $[Al_6(AsO_4)_3(OH)_9(H_2O)_4]\cdot 2H_2O$	1	1.05	105	500 ms	$3\ e^-/Å^2$

3. Results

In this manuscript we present the results on the structure solution and refinement of four compounds. The first compound, $Sr_5CuGe_9O_{24}$, is not beam sensitive and a low-dose technique is not really necessary for diffraction experiments on this crystal. Its rather complex structure for an oxide with 22 independent atom positions [20] and good crystalline quality makes it a good test structure for our method in order to determine the minimal dose necessary for structure solution and refinement.

The second compound Mn-formiate ($[Mn(HCOO)_2(H_2O)_2]_\infty$, is a metal-organic framework (MOF) structure [21]. This MOF is known for its beam sensitivity and even though its structure is not very complex the solution and refinement of this structure is a real application of LD-EDT.

The third compound was a new germanium-based garnet-like structure containing Na atoms and presenting a site with mixed Na/Mn occupancy. In this case we were able to refine the Na/Mn ratio.

Finally, LD-EDT has been applied to solve a previously unknown mineral structure, bulachite $[Al_6(AsO_4)_3(OH)_9(H_2O)_4]\cdot 2H_2O$ [22]. This mineral contains unbound water molecules which make it beam sensitive.

3.1. Structure Solution

The program PETS2 yields the unit cell parameters and the sections of the reconstructed 3-dimensional reciprocal lattice. From the systematic extinctions observed on these sections the possible space groups can be deduced. The structure solution is then attempted using the SIR2014 and Superflip programs embedded into the Jana2006 suite.

3.1.1. $Sr_5CuGe_9O_{24}$

The unit cell was determined to be monoclinic with cell parameters a = 12.20 Å, b = 8.39 Å, c = 10.57 Å and β = 101.6°. Figure 1 shows the sections $hk0$, $h0l$, and $0kl$ of the reconstructed 3D reciprocal lattice and the cylindrical projection of the intensity clusters used for the refinement of the tilt axis orientation. From the systematic absences (hkl: none, $h0l$: $l = 2n$) the possible space groups Pc and $P2/c$ were deduced. The best structure solution results were obtained by SIR2014 in space group $P2/c$. The data set contains 1775 independent reflections representing a completeness of 69% up to a resolution of 0.8 Å.

The structure solution obtained by SIR2014 with a final residual value R = 30 % contained all atom positions (cf. Table 2). The three positions with the highest peaks in the potential map correspond to the heaviest ions (Sr), followed by the 5 positions for Ge and the single position for Cu. The next 13 positions are indeed the correct positions for the oxygen ions. The following 3 positions also interpreted as oxygen by SIR2014 have significantly lower peak heights in the potential map and are spurious peaks. The projection along the b-axis of the resulting structure is shown in Figure 2.

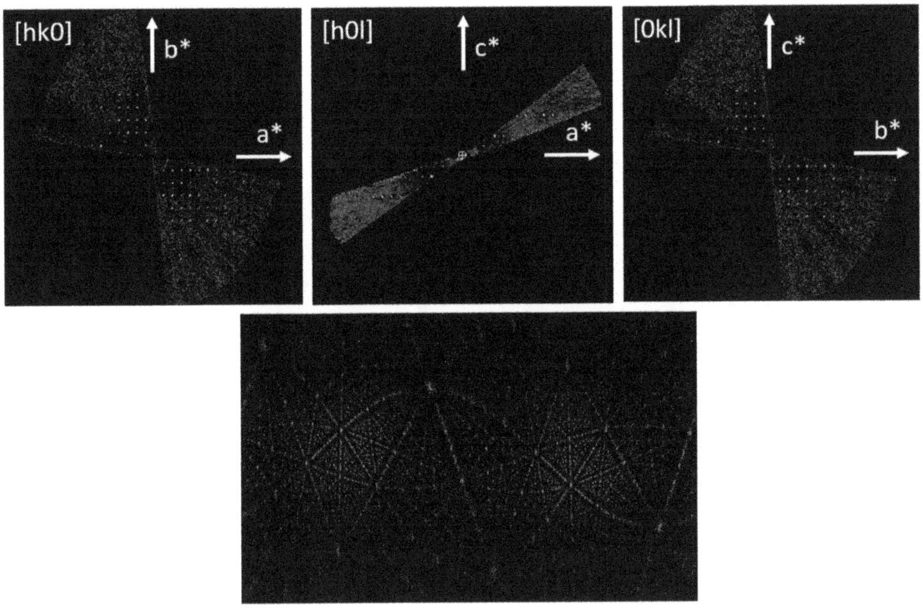

Figure 1. Sections of the reconstructed 3D reciprocal lattice of $Sr_5CuGe_9O_{24}$ (**top**), cylindrical projection of the intensity clusters used for refining the tilt axis orientation (**bottom**).

Table 2. Structure solution of $Sr_5CuGe_9O_{24}$ obtained by electron diffraction and comparison to the positions obtained by X-ray diffraction [20].

Atom	Wyck	Electron Diffraction Structure Solution			X-ray Diffraction				Difference (Å)	Average Difference (Å)
		x	y	z	x	y	z	Uiso		
Sr1	2e	0	0.357	0.75	0	0.3531(8)	0.75	0.007(2)	0.031	
Sr2	4g	0.757	0.030	0.684	0.7559(4)	0.0316(5)	0.6837(4)	0.0047(18)	0.018	
Sr3	4g	0.652	0.483	0.680	0.6505(4)	0.4849(5)	0.6817(5)	0.0080(17)	0.031	
Ge1	4g	0.852	0.709	0.936	0.8529(5)	0.7110(6)	0.9356(5)	0.0054(19)	0.020	
Ge2	4g	0.589	0.174	0.371	0.5889(4)	0.1750(7)	0.3707(6)	0.012(2)	0.008	
Ge3	4g	0.788	0.669	0.489	0.7896(5)	0.6712(7)	0.4765(5)	0.0072(18)	0.039	
Ge4	4g	0.540	0.844	0.411	0.5412(4)	0.8401(7)	0.4122(6)	0.009(2)	0.036	
Ge5	2e	0	0.780	0.75	0	0.7801(9)	0.75	0.008(2)	0.001	
Cu1	2a	0	0	0.5	0	0	0	0.006(3)	0.000	
Cations										0.021
O1	4g	0.808	0.690	0.310	0.797(2)	0.695(3)	0.315(3)	0.011(2)	0.159	
O2	4g	0.542	0.958	0.626	0.544(2)	0.964(3)	0.630(3)	0.011(2)	0.063	
O3	4g	0.922	0.607	0.841	0.944(2)	0.593(3)	0.827(3)	0.011(2)	0.336	
O4	4g	0.644	0.677	0.496	0.646(2)	0.675(3)	0.489(3)	0.011(2)	0.088	
O5	4g	0.409	0.749	0.473	0.415(2)	0.736(3)	0.469(3)	0.011(2)	0.141	
O6	4g	0.662	0.985	0.410	0.671(2)	0.977(3)	0.421(3)	0.011(2)	0.150	
O7	2f	0.5	0.307	0.25	0.5	0.344(5)	0.25	0.011(2)	0.303	
O8	4g	0.899	0.841	0.062	0.910(2)	0.818(3)	0.065(3)	0.011(2)	0.224	
O9	4g	0.861	0.798	0.598	0.869(2)	0.784(3)	0.595(3)	0.011(2)	0.158	
O10	4g	0.704	0.733	0.828	0.716(2)	0.724(3)	0.836(3)	0.011(2)	0.171	
O11	4g	0.823	0.535	0.053	0.817(2)	0.527(3)	0.026(2)	0.011(2)	0.281	
O12	2f	0.5	0.755	0.25	0.5	0.740(4)	0.25	0.011(2)	0.122	
O13	4g	0.913	0.911	0.831	0.909(2)	0.921(3)	0.820(3)	0.011(2)	0.140	
Oxygens										0.180

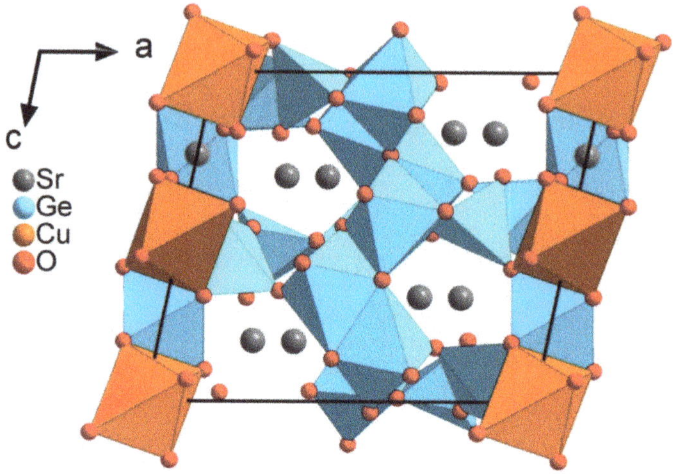

Figure 2. Projection along the b-axis of the structure of $Sr_5CuGe_9O_{24}$.

Comparing these results with the atom positions refined from X-ray diffraction, one can see in Table 2 that the average distance for the cations is 0.027 Å and for the oxygen it is 0.173 Å. The high accuracy of the positions obtained in the structure solution indicates already the high quality of the diffraction data.

3.1.2. Mn-Formiate ($[Mn(HCOO)_2(H_2O)_2]_\infty$

The unit cell was determined to be monoclinic with cell parameters a = 8.8 Å, b = 7.2 Å, c = 9.6 Å, and β = 97.7°. Figure 3 shows the sections *hk*0, *h*0*l*, and 0*kl* of the reconstructed 3D reciprocal lattice and the cylindrical projection of the intensity clusters used for the refinement of the tilt axis orientation. The systematic absences were compatible with the known space group $P2_1/c$. The best structure solution results were obtained by SIR2014 in space group $P2_1/c$. The data set contains 976 independent reflections representing a completeness of 62% up to a resolution of 0.8 Å.

The structure solution obtained by SIR2014 with a final residual value R = 34 % contained all atom positions (cf. Table 3). SIR2014 correctly assigned Mn to the two highest peaks in the potential map. The C and O atoms being neighbors in the periodic table it is very difficult to distinguish between them on the basis of scattering power. Most of these atoms were also correctly assigned, except for O6 and C2, which were inversed. Knowing the molecule used in the synthesis, the correct identification of carbon and oxygen is no problem. The projection along the b-axis of the resulting structure is shown in Figure 4.

The Mn are on special positions of the space group and the coordinates are therefore fixed by symmetry and cannot differ from those obtained from X-ray data. The coordinates of the O and C, however, are not fixed and the comparison shows that the average distance of the found positions is only 0.118 Å (Table 3) with respect to the atom positions refined against X-ray diffraction data. As for the $Sr_5CuGe_9O_{24}$, the accuracy of the positions shows once again the high quality of the data.

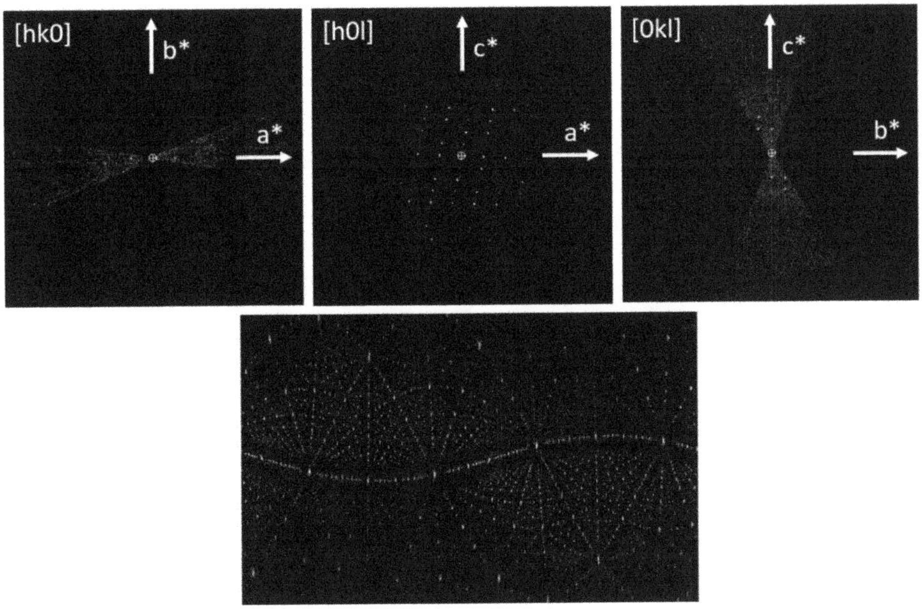

Figure 3. Sections of the reconstructed 3D reciprocal lattice of Mn-formiate (**top**), cylindrical projection of the intensity clusters used for refining the tilt axis orientation (**bottom**).

Table 3. Structure solution of Mn-formiate obtained by electron diffraction and comparison to the positions obtained by X-ray diffraction [21].

Atom	Wyck	Electron Diffraction Structure Solution			X-ray Diffraction				Difference (Å)	Average Difference (Å)
		x	y	z	x	y	z	Uiso		
Mn1	2a	0	0	0	0	0	0	0.004	0.000	
Mn2	2d	0.5	0.5	0	0.5	0.5	0	0.005	0.000	
Mn										0
O1	4e	0.773	0.143	0.003	0.78450(5)	0.15120(7)	0.00599(5)	0.008	0.154	
O2	4e	0.563	0.217	0.081	0.56126(6)	0.21693(8)	0.08061(6)	0.008	0.024	
O3	4e	0.905	0.597	0.295	0.90332(6)	0.60127(8)	0.29393(5)	0.007	0.036	
O4	4e	0.920	0.769	0.105	0.91588(6)	0.75836(7)	0.09698(5)	0.007	0.116	
O5	4e	0.588	0.593	0.197	0.59155(6)	0.61552(10)	0.20059(6)	0.01	0.197	
O6	4e	0.265	0.467	0.061	0.27190(5)	0.48558(8)	0.06818(5)	0.008	0.184	
C1	4e	0.328	0.897	0.939	0.67129(5)	0.11072(7)	0.06784(5)	0.007	0.100	
C2	4e	0.955	0.722	0.223	0.96723(6)	0.71297(7)	0.22071(5)	0.008	0.130	
O and C										0.118

3.1.3. Garnet $(Na_{2/3}Mn_{1/3})_3Ge_2Ge_3O_{12}$

In the framework of expanding the range of germanium-based compounds in the pyroxene family, a sample with the nominal composition $NaMnGe_2O_6$ was synthesized at high temperature and high pressure. X-ray powder diffraction (XRPD) data were collected from $2\theta = 16°$ to $60°$ using a D5000 Bruker diffractometer equipped with a focusing Ge(111) primary monochromator and an Elphyse linear detector, revealing the multi-phase nature of the sample. Examination of the diffraction pattern revealed the presence of the mixed Na, Mn germanate $Na_2(Mn_{0.26}\ Na_{0.74})\ Mn_5Ge_6O_{20}$ [23], the remaining peaks being indexable in a body-centered cubic unit cell with a $\approx$12 Å, similar to that of garnets such as e.g., $CdGeO_3$ [24]. A successful Rietveld refinement was performed using the Fullprof

Suite software [25] based on the garnet structure in the cubic *Ia-3d* (n° 230) space group symmetry. Ge atoms were placed at the *16a* and *24d* Wyckoff positions, a fixed composition $Mn_{1/3}Na_{2/3}$ mixture at the *24c* position, according to the stoichiometry deduced from the EDX measurements, and the oxygen atom at the *96c* position. The scale factor, cell parameter, oxygen atom position, and an overall atomic displacement parameter (a.d.p.) were refined. The reflection profiles for both phases were treated using pseudo-Voigt functions. The atomic parameters of the secondary phase were taken from literature and fixed; only an overall a.d.p., profile and cell parameters, and scale factor were refined. The refinement yielded $Rwp = 15.2\%$, $R_{Bragg} = 3.44\%$, $\chi^2 = 6.76$ with cell parameter $a = 11.9860(3)$ Å and a mass fractions of 64(1)% for the garnet phase (Figure 5). Given the high impurity phase proportion of the sample, it was decided to reinvestigate the structure of this new garnet compound using electron diffraction.

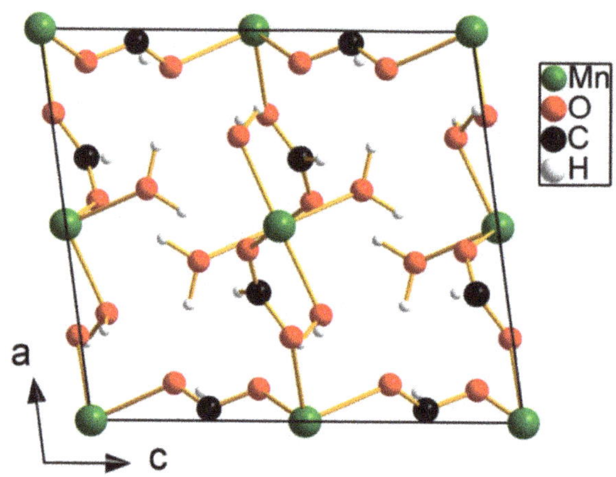

Figure 4. Projection along the b-axis of the structure of Mn-formiate ($[Mn(HCOO)_2(H_2O)_2]_\infty$.

According to our experience, compounds containing alkaline species are most time sensitive to irradiation, therefore we decided to test the LD-EDT approach on this new garnet-like compound. The chemical composition was determined by EDX to be 20% Na, 11% Mn, and 69% Ge. LD-EDT data led to a cubic unit cell with parameter $a = 12.2$ Å. Figure 6 shows the sections *hk*0, *h*0*l*, and 0*kl* of the reconstructed 3D reciprocal lattice and the cylindrical projection of the intensity clusters used for the refinement of the tilt axis orientation. The structure was solved with a final *R*-value equal to 28% by Superflip from a data set containing 220 independent reflections representing a completeness of 98% up to a resolution of 0.7 Å. The proposed space group is *Ia-3d (n° 230)*. The structure agrees with the XRPD results reported above. The new garnet-like compound presents a very high symmetry, and only four independent positions. Atomic attribution from Superflip was corrected using chemical information from EDX composition and interatomic distances and the resulting structure is shown in Figure 7.

3.1.4. Bulachite $[Al_6(AsO_4)_3(OH)_9(H_2O)_4]\cdot 2H_2O$

The orthorhombic unit cell was determined by PETS to have cell parameters of $a = 15.40$ Å, $b = 17.66$ Å, and $c = 7.81$ Å. Figure 8 shows the sections *hk*0, *h*0*l*, and 0*kl* of the reconstructed 3D reciprocal lattice and the cylindrical projection of the intensity clusters used for the refinement of the tilt axis orientation. The systematic extinctions were compatible with space groups $Pn2_1a$ and *Pnma*. The data set contains 2438 independent reflections which corresponds to a completeness of 80% up to 0.8 Å resolution.

The best structure solution was obtained by SIR2014 in space group *Pnma* with a final *R* value of 30%. All cation positions were obtained, attributed by SIR2014 to elements according to their number of electrons: the first position was attributed to As, followed by the Al. The next 14 positions correspond to the oxygen forming the coordination polyhedra around the cations. The following two positions that SIR2014 interpreted as being hydrogen correspond in fact to the oxygen positions of free water molecules in between the polyhedral layers (see Table 4 and Figure 9). This structure solution was used for the refinement of synchrotron powder diffraction data, which confirmed the structure and allowed identifying oxygen atoms participating to hydroxyl groups, bonded and free water molecules among the found oxygen atoms using BVS calculations [22]. Table 4 shows that the average distance between the cation positions in the unrefined structural model found by LD-EDT and the positions from Rietveld refinement [22] is 0.079 Å, for the O^{2-} and OH^- anions the average distance is 0.110 Å, while for the oxygen positions from bonded H_2O molecules the distance is 0.186 Å. For the also detected free water molecules in between the polyhedral layers the precision is less good with an average distance of 0.324 Å.

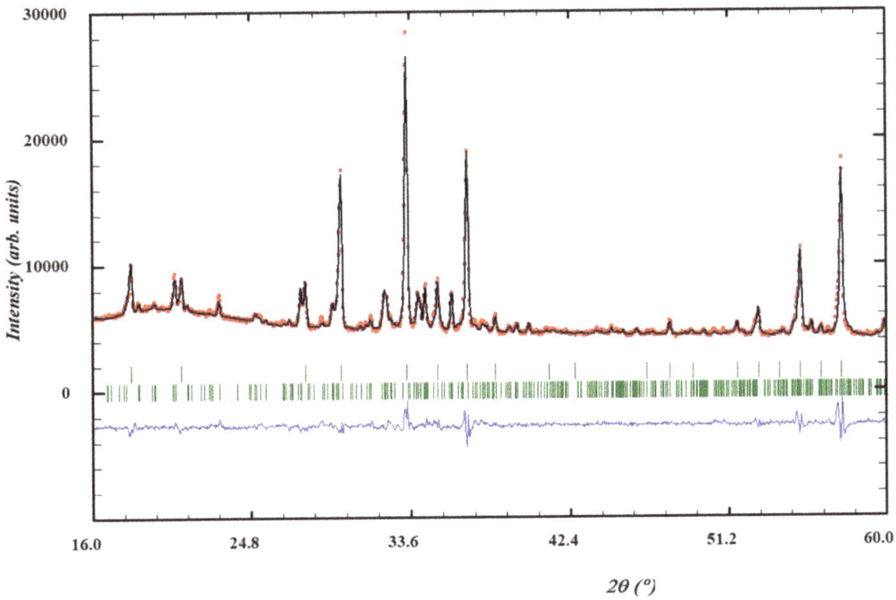

Figure 5. Rietveld plot for the refinement of the $(Na_{2/3}Mn_{1/3})_3Ge_2Ge_3O_{12}$ garnet structure. Red dots: observed data, black line calculated pattern, blue line: difference. The tick marks are for $(Na_{2/3}Mn_{1/3})_3Ge_2Ge_3O_{12}$ top and $Na_2(Mn_{5.26}Na_{0.74})Ge_6O_{20}$ (bottom).

3.2. *Structure Refinement*

After structure solution, first a kinematical refinement was performed in Jana2006 using the standard settings of all parameters. This refinement usually improved the *R*-values slightly but no major changes in atom positions were observed. The structure refinement based on the dynamical theory of diffraction, however, significantly improved the atom positions and reduced the differences to the X-ray refined positions. The dynamical refinement was also performed in the Jana2006 suite. For this refinement several parameters can be chosen. For all structures presented here we first ran the refinement using default values of all parameters (observed reflections: $I/\sigma(I) > 3$, maximal diffraction vector g(max) = 2 $Å^{-1}$, maximal excitation error (Matrix) = 0.01 $Å^{-1}$, maximal excitation error (Refine) = 0.1 $Å^{-1}$, RSg(max) = 0.4, number of integration steps = 128). In each case the refinement converged to reasonable *R* values and atomic positions and/or occupancies. An improvement of the results could then be obtained in some cases by lowering the $I/\sigma(I)$ limit for observed reflections. Indeed, the TVIPS F416 camera has a high sensitivity (85 counts per incident electron) and a low noise level (variance between 2 and 5 for exposure times up to 1 s). These characteristics allowed us reducing the limit for observed reflections $I/\sigma(I)$ from the default value of 3 to values as low as 1.2. This increased the number of reflections used in the refinements and resulted in improved atomic positions in the final solutions. The final *R*-values and GoF of the refined structures are reported in Table 5.

Table 4. Structure solution of bulachite obtained by electron diffraction and comparison to the positions obtained by X-ray diffraction [22].

Atom	Wyck	Electron Diffraction Structure Solution			Synchrotron Radiation Diffraction			Difference (Å)
		x	y	z	x	Y	z	
As1	8d	0.825	0.010	0.383	0.82245(18)	0.00921(19)	0.3827(4)	0.042
As2	4c	0.900	0.25	0.574	0.9057(3)	0.25	0.5824(6)	0.111
Al1	8d	0.634	0.070	0.242	0.6329(6)	0.0696(5)	0.2444(17)	0.027
Al2	8d	0.837	0.100	0.752	0.8397(6)	0.0964(5)	0.7535(16)	0.064
Al3	8d	0.711	0.169	0.495	0.7207(6)	0.1692(6)	0.4971(16)	0.152
O1	8d	0.733	0.998	0.284	0.7276(5)	0.9939(6)	0.2856(14)	0.112
O2	8d	0.895	0.019	0.230	0.9002(6)	0.0190(6)	0.2359(12)	0.094
O3	8d	0.815	0.092	0.500	0.8160(7)	0.0885(5)	0.5006(13)	0.064
O4	8d	0.846	0.936	0.503	0.8459(7)	0.9358(5)	0.5099(13)	0.055
O5	4c	0.802	0.25	0.495	0.8043(5)	0.25	0.5078(18)	0.108
O6	4c	0.984	0.25	0.418	0.9752(8)	0.25	0.4186(14)	0.137
O7	8d	0.917	0.175	0.701	0.9217(6)	0.17267(14)	0.7019(7)	0.084
Oh1	8d	0.708	0.158	0.250	0.7085(9)	0.1522(7)	0.250(2)	0.103
Oh2	8d	0.639	0.095	0.480	0.6300(10)	0.0952(9)	0.4819(19)	0.141
Oh3	8d	0.919	0.021	0.749	0.9259(8)	0.0208(7)	0.772(3)	0.212
Oh4	8d	0.739	0.157	0.735	0.7386(8)	0.1541(7)	0.740(2)	0.065
Oh5	4c	0.646	0.25	0.522	0.6390(13)	0.25	0.511(3)	0.140
Ow1	8d	0.518	0.121	0.202	0.5229(9)	0.1135(7)	0.226(3)	0.245
Ow2	8d	0.848	0.116	0.991	0.8415(13)	0.1119(9)	0.9944(19)	0.128
W1	4c	0.948	0.25	0.029	0.93	0.25	0.018	0.294
W2	4c	0.363	0.75	0.063	0.3652	0.75	0.0187	0.354

Table 5. Final *R*-values of the dynamical refinements.

	R(obs)	wR(obs)	R(all)	wR(all)	GoF(all)	GoF(obs)
$Sr_5CuGe_9O_{24}$	5.80	6.87	33.69	14.65	0.84	0.41
Mn-formiate $[Mn(HCOO)_2(H_2O)_2]_\infty$	7.25	9.53	58.66	18.48	1.29	0.43
Garnet $(Na_{2/3}Mn_{1/3})_3Ge_2Ge_3O_{12}$	7.62	9.15	44.80	17.02	0.96	0.48
Bulachite $[Al_6(AsO_4)_3(OH)_9(H_2O)_4]\cdot 2H_2O$	10.56	11.15	24.68	13.37	2.24	1.24

3.2.1. $Sr_5CuGe_9O_{24}$

Starting from the atomic positions obtained from the structure solution the coordinates of the atoms and their isotropic thermal displacement parameters were refined. The value of the thickness of the crystal was refined to 81.1 nm.

While the precision of the cation positions did not change much, the precision of the oxygen positions was improved from an average distance to the X-ray refined positions of 0.180 Å to 0.152 Å by the refinement (cf. Table 6).

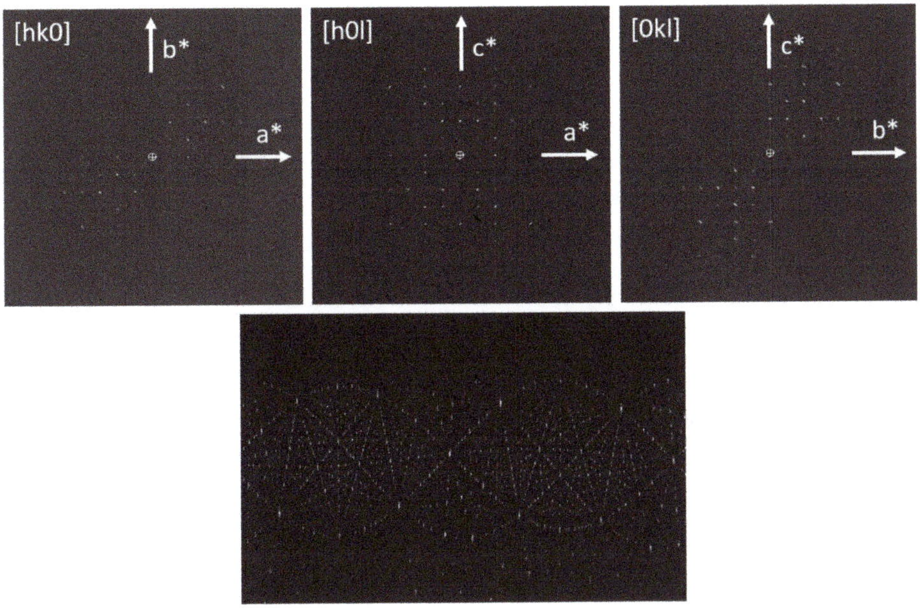

Figure 6. Sections of the reconstructed 3D reciprocal lattice of the garnet $(Na_{2/3}Mn_{1/3})_3Ge_2Ge_3O_{12}$ (**top**), cylindrical projection of the intensity clusters used for refining the tilt axis orientation (**bottom**).

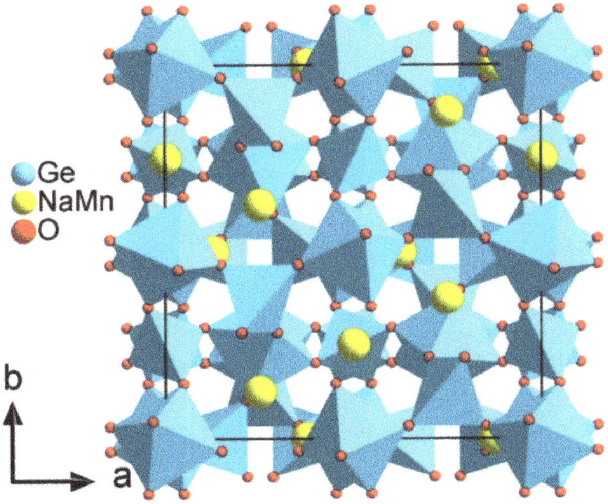

Figure 7. Projection along the c-axis of the structure of the garnet $(Na_{2/3}Mn_{1/3})_3Ge_2Ge_3O_{12}$.

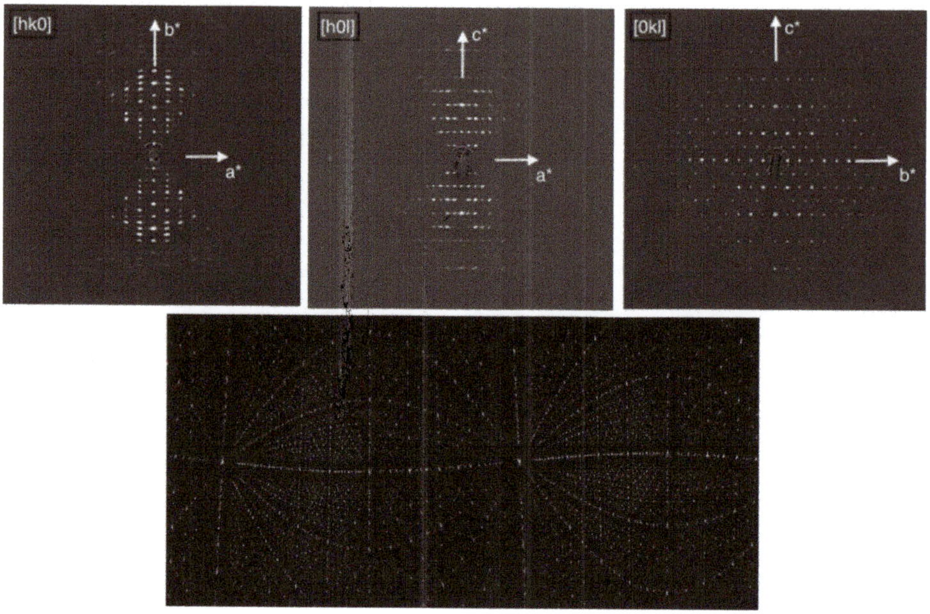

Figure 8. Sections of the reconstructed 3D reciprocal lattice of bulachite (**top**), cylindrical projection of the intensity clusters used for refining the tilt axis orientation (**bottom**).

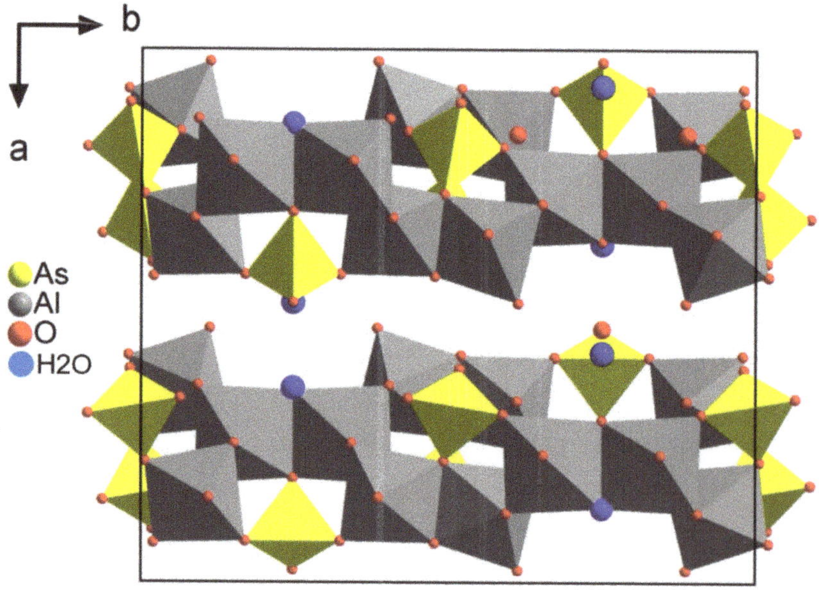

Figure 9. Projection along the c-axis of the structure of the mineral bulachite $[Al_6(AsO_4)_3(OH)_9(H_2O)_4]\cdot 2H_2O$.

Table 6. Comparison of the atomic positions for $Sr_5CuGe_9O_{24}$ refined using electron diffraction data with those refined using X-ray diffraction data.

Atom	Wyck	X-ray Diffraction				Electron Diffraction Dynamical Refinement				Difference (Å)	Average Difference (Å)
		x	y	z	Uiso	x	y	z	Uiso		
Sr1	2e	0	0.3531(8)	0.75	0.007(2)	0	0.3561(6)	0.75	−0.0179(3)	0.024	
Sr2	4g	0.7559(4)	0.0316(5)	0.6837(4)	0.0047(18)	0.7553(5)	0.0322(6)	0.6833(6)	−0.0039(11)	0.009	
Sr3	4g	0.6505(4)	0.4849(5)	0.6817(5)	0.0080(17)	0.6520(4)	0.4850(6)	0.6814(5)	−0.0110(7)	0.019	
Ge1	4g	0.8529(5)	0.7110(6)	0.9356(5)	0.0054(19)	0.8524(3)	0.7086(4)	0.9339(4)	−0.0192(2)	0.026	
Ge2	4g	0.5889(4)	0.1750(7)	0.3707(6)	0.012(2)	0.5865(6)	0.1747(8)	0.3711(8)	−0.0202(2)	0.029	
Ge3	4g	0.7896(5)	0.6712(7)	0.4765(5)	0.0072(18)	0.7859(3)	0.6706(4)	0.4772(4)	−0.0200(3)	0.045	
Ge4	4g	0.5412(4)	0.8401(7)	0.4122(6)	0.009(2)	0.5403(7)	0.8464(9)	0.4120(9)	0.0033(15)	0.052	
Ge5	2e	0	0.7801(9)	0.75	0.008(2)	0	0.7817(5)	0.75	0.0008(14)	0.013	
Cu1	2a	0	0	0	0.006(3)	0	0	0.5	0.0037(22)	0.000	
Cations											0.024
O1	4g	0.797(2)	0.695(3)	0.315(3)	0.011(2)	0.7981(13)	0.6906(18)	0.3083(16)	−0.0076(27)	0.080	
O2	4g	0.544(2)	0.964(3)	0.630(3)	0.011(2)	0.5411(9)	0.9594(12)	0.6305(12)	−0.0178(9)	0.052	
O3	4g	0.944(2)	0.593(3)	0.827(3)	0.011(2)	0.9390(14)	0.6108(21)	0.8369(19)	−0.0051(32)	0.186	
O4	4g	0.646(2)	0.675(3)	0.489(3)	0.011(2)	0.6491(6)	0.6840(8)	0.4929(7)	−0.0209(4)	0.129	
O5	4g	0.415(2)	0.736(3)	0.469(3)	0.011(2)	0.4115(25)	0.7426(31)	0.4586(33)	0.0442(81)	0.116	
O6	4g	0.671(2)	0.977(3)	0.421(3)	0.011(2)	0.6619(16)	0.9801(22)	0.4034(21)	0.0057(44)	0.189	
O7	2f	0.5	0.344(5)	0.25	0.011(2)	0.5	0.3095(14)	0.25	−0.0190(9)	0.283	
O8	4g	0.910(2)	0.818(3)	0.065(3)	0.011(2)	0.9003(14)	0.8253(20)	0.0713(19)	−0.0028(34)	0.150	
O9	4g	0.869(2)	0.784(3)	0.595(3)	0.011(2)	0.8685(18)	0.8045(25)	0.6051(23)	0.0154(53)	0.198	
O10	4g	0.716(2)	0.724(3)	0.836(3)	0.011(2)	0.7151(16)	0.7362(22)	0.8340(22)	0.0071(47)	0.104	
O11	4g	0.817(2)	0.527(3)	0.026(2)	0.011(2)	0.8280(19)	0.5366(25)	0.0305(23)	0.0141(53)	0.157	
O12	2f	0.5	0.740(4)	0.25	0.011(2)	0.5	0.7521(43)	0.25	0.0402(112)	0.098	
O13	4g	0.909(2)	0.921(3)	0.820(3)	0.011(2)	0.9184(13)	0.9043(18)	0.8380(17)	−0.0101(24)	0.238	
Oxygens											0.152

The refined structure of $Sr_5CuGe_9O_{24}$ confirms the particular features of this compound where three different environments for Ge ions with four-fold, five-fold, and six-fold coordination are present, respectively (Figure 2). The Cu cations are coordinated by oxygen octahedra.

3.2.2. Mn-Formiate

The dynamical refinement of the atomic coordinates and the isotropic thermal displacement parameters of Mn-formiate yielded the positions in Table 7. The value of the thickness of the crystal was refined to 54.1 nm.

In this particular case, the initial structure obtained by SIR2014 was already very close to the X-ray refined structure, so that the refinement even though it converged easily, did not make the model more accurate.

3.2.3. Garnet $(Na_{2/3}Mn_{1/3})_3Ge_2Ge_3O_{12}$

The cations being on special positions, there is no degree of freedom in the coordinates and the positions are identical to those in the X-ray refined structure. The O position, however, is not fixed and one cation position has mixed Na/Mn occupancy. We applied a dynamical refinement using PETS2 and JANA2006 to refine the Na/Mn occupancy and the oxygen position from the LD-EDT data. The value of the thickness of the crystal was refined to 61.6 nm. The precision of the oxygen position was improved from a distance to the X-ray refined position of 0.121 Å to 0.046 Å by the refinement. The Na/Mn ratio initially fixed to 0.5/0.5 converged rapidly to 0.64/0.36, very close to the X-ray refinement result. (Table 8).

It can be noted that a kinematical refinement of the Na/Mn ratio on the same data is not stable.

These results lead to two conclusions: First, the irradiation during the LD-EDT did not modify the composition of the compound through beam damage such as radiolysis; second, the good quality of the data allowed refining the occupancy ratio of the mixed site.

3.2.4. Bulachite

In the first steps of the refinement the atom positions identified by SIR2014 as H1 and H2 were not retained. A Fourier difference map, however, confirmed the existence of these ion positions and suggested that they are indeed oxygen positions of non-bonded water molecules. Including these positions and refining the coordinates and isotropic atomic displacement parameters of all atoms lead to final R values as presented in Table 5. The value of the thickness of the crystal was refined to 50.7 nm. It can be noted that the thermal displacement parameters of all atoms could be refined to meaningful values. The values of U_{iso} of the non-bonded water molecules is significantly higher than those of the other oxygen ions.

The atomic coordinates of the refined structure are presented in Table 9. There is an improvement in the precision of the cations and the O^{2-} and OH^{-} positions, but the distances of the water molecules from the X-ray refined positions seem to increase slightly.

Table 7. Comparison of the atomic positions for Mn-formiate refined using electron diffraction data with those refined using X-ray diffraction data [21].

Atom	Wyck	X-ray Diffraction				Electron Diffraction Refinement				Difference (Å)	Average Difference (Å)
		x	y	z	Uiso	x	y	z	Uiso		
Mn1	2a	0	0	0	0.004	0	0	0	−0.0208(6)	0.000	
Mn2	2d	0.5	0.5	0	0.005	0.5	0.5	0	0.0190(20)	0.000	
Mn											0
O1	4e	0.78450(5)	0.15120(7)	0.00599(5)	0.008	0.7822(12)	0.1657(24)	0.0020(11)	0.0132(31)	0.118	
O2	4e	0.56126(6)	0.21693(8)	0.08061(6)	0.008	0.5811(10)	0.1926(22)	0.0769(9)	−0.0006(24)	0.263	
O3	4e	0.90332(6)	0.60127(8)	0.29393(5)	0.007	0.9102(13)	0.6294(25)	0.2995(13)	0.0117(32)	0.251	
O4	4e	0.91588(6)	0.75836(7)	0.09698(5)	0.007	0.937(11)	0.7687(25)	0.1142(10)	0.0112(29)	0.246	
O5	4e	0.59155(6)	0.61552(10)	0.20059(6)	0.01	0.5930(16)	0.6017(35)	0.2174(17)	0.0700(62)	0.193	
O6	4e	0.27190(5)	0.48558(8)	0.06818(5)	0.008	0.2667(13)	0.4926(32)	0.0573(14)	0.0289(40)	0.120	
C1	4e	0.67129(5)	0.11072(7)	0.06784(5)	0.007	0.6775(6)	0.1198(11)	0.0636(5)	−0.0311(4)	0.101	
C2	4e	0.96723(6)	0.71297(7)	0.22071(5)	0.008	0.9790(12)	0.7221(28)	0.2247(11)	0.0120(32)	0.131	
O and C											0.178

Table 8. Comparison of the refined atomic positions and occupancies from LD-EDT data for the garnet $(Na_{2/3}Mn_{1/3})_3Ge_2Ge_3O_{12}$ with the X-ray refined positions.

Atom	Wyck	Dynamical Refinement From LD-EDT Data					X-ray Diffraction Refinement					Distance (Å)
		x	y	z	Occ	Uiso	x	y	z	Occ	Uiso	
Ge1	16a	0	0	0	1	−0.0215(1)	0	0	0	1	0.01267	
Ge2	24d	3/8	0	1/4	1	0.0044(8)	3/8	0	1/4	1	0.00633	
Mn1	24c	1/8	0	1/4	0.36(2)	0.021(2)	1/8	0	1/4	1/3	0.00887	
Na1	24c	1/8	0	1/4	0.64(2)	0.021(2)	1/8	0	1/4	2/3	0.00887	
O1	96h	0.0339(4)	0.0471(4)	0.6474(4)	1	0.0008(9)	0.0301(6)	0.0469(7)	0.6470(7)	1	0.01267	0.046

Table 9. Comparison of the atomic positions for bulachite refined using electron diffraction data with those refined using X-ray diffraction data.

Atom	Wyck	Synchrotron Radiation Diffraction				Electron Diffraction Refinement				Difference (Å)	
		x	y	z	B (Å^2)	x	y	z	Uiso		
As1	8d	0.82245(18)	0.00921(19)	0.3827(4)	0.0368(6)	0.8242(2)	0.0097(1)	0.3832(2)	0.0086(4)	0.031	
As2	4c	0.9057(3)	0.25	0.5824(6)	0.0368(6)	0.9034(3)	0.25	0.5773(3)	0.0074(6)	0.054	
Al1	8d	0.6329(6)	0.0696(5)	0.2444(17)	0.0368(6)	0.6334(4)	0.0720(2)	0.2396(4)	0.0051(8)	0.055	**Average cation distance: 0.057 Å**
Al2	8d	0.8397(6)	0.0964(5)	0.7535(16)	0.0368(6)	0.8393(4)	0.0966(2)	0.7480(4)	0.0063(7)	0.056	
Al3	8d	0.7207(6)	0.1692(6)	0.4971(16)	0.0368(6)	0.7140(4)	0.1665(2)	0.4916(4)	0.0052(8)	0.091	
O1	8d	0.7276(5)	0.9939(6)	0.2856(14)	0.0186(15)	0.7289(7)	0.9977(3)	0.2846(6)	0.0223(13)	0.071	
O2	8d	0.9002(6)	0.0190(6)	0.2359(12)	0.0186(15)	0.9011(7)	0.0219(3)	0.2307(6)	0.0244(13)	0.067	
O3	8d	0.8160(7)	0.0885(5)	0.5006(13)	0.0186(15)	0.8128(7)	0.0906(3)	0.4958(6)	0.0241(13)	0.073	
O4	8d	0.8459(7)	0.9358(5)	0.5099(13)	0.0186(15)	0.8407(7)	0.9334(3)	0.5053(6)	0.0246(13)	0.099	
O5	4c	0.8043(5)	0.25	0.5078(18)	0.0186(15)	0.8037(8)	0.25	0.4855(7)	0.0060(14)	0.178	**Average oxygen distance (with ionic bonding): 0.102Å**
O6	4c	0.9752(8)	0.25	0.4186(14)	0.0186(15)	0.9862(11)	0.25	0.4348(10)	0.0380(24)	0.191	
O7	8d	0.9217(6)	0.17267(14)	0.7019(7)	0.0186(15)	0.9179(7)	0.1751(3)	0.7036(6)	0.0149(12)	0.075	
Oh1	8d	0.7085(9)	0.1522(7)	0.250(2)	0.0186(15)	0.7088(6)	0.1552(2)	0.2458(6)	0.0107(11)	0.062	
Oh2	8d	0.6300(10)	0.0952(9)	0.4819(19)	0.0186(15)	0.6295(7)	0.0921(3)	0.4776(6)	0.0200(13)	0.050	
Oh3	8d	0.9259(8)	0.0208(7)	0.772(3)	0.0186(15)	0.9323(6)	0.0219(2)	0.7482(6)	0.0107(11)	0.215	
Oh4	8d	0.7386(8)	0.1541(7)	0.740(2)	0.0186(15)	0.7368(6)	0.1569(2)	0.7322(6)	0.0167(12)	0.084	
Oh5	4c	0.6390(13)	0.25	0.511(3)	0.0186(15)	0.6445(9)	0.25	0.5121(7)	0.0084(15)	0.064	
Ow1	8d	0.5229(9)	0.1135(7)	0.226(3)	0.0186(15)	0.5233(8)	0.1238(3)	0.2010(7)	0.0346(16)	0.270	**H_2O polyhedra: 0.198 Å**
Ow2	8d	0.8415(13)	0.1119(9)	0.9944(19)	0.0186(15)	0.8420(8)	0.1189(3)	0.9924(7)	0.0341(15)	0.125	
W1	4c	0.93	0.25	0.018	0.074(11)	0.9654(13)	0.25	0.0290(12)	0.0611(31)	0.560	
W2	4c	0.3652	0.75	0.0187	0.074(11)	0.3400(20)	0.75	0.0581(18)	0.1872(66)	0.503	**Free H_2O: 0.53 Å**

4. Discussion and Conclusions

Working with beam sensitive materials, reducing the incident electron dose on the crystal is paramount in order to study its structure without causing structural changes during the experiment. Exposing the crystal to the incident beam exclusively during the recording of the diffracted intensities and blanking the beam at any other time seems to be the obvious basis.

The subsequent step is to optimize the signal to noise ratio of the diffracted intensities. This can be achieved by increasing the signal and/or reducing the noise. For a given illumination condition, i.e., a given dose rate, choosing a relatively large crystal allows obtaining higher diffracted intensities. This is possible in LD-EDT where the size limit for the crystal is given by the condition that it should stay inside the selected area aperture during the entire tomography. Using an aperture that selects an area of about 2 μm in diameter, a crystal of the order of 1 μm can easily be chosen. The large selected area aperture can lead to difficulties if the crystal is bent or if it contains smaller crystalline domains. In this case, choosing a different particle usually solves the problem.

Reducing the noise of the acquisition largely depends on the camera used for recording the diffraction frames. Modern direct detection cameras certainly are a good albeit costly choice for low noise acquisition. In this work we have shown that a sensitive CMOS camera also qualifies for the necessary signal to noise ratio that allows accurate data acquisition at a dose as low as $0.1\ e^-/\text{Å}^2$ for the entire data set.

Of course, using such a low dose will result in many reflections being very weak. The usual criteria for counting a reflection as observed ($I/\sigma(I) > 3$) will eliminate many reflections, leaving a small data set that may not be sufficient for the solution and refinement of complex structures. In the case of $Sr_5CuGe_9O_{24}$ and due to the extremely low dose used for the data acquisition, we had to change this criterion and consider all reflections with $I/\sigma(I) > 1.2$ as observed. This was possible because of the low background noise of the TVIPS F416 camera and it was sufficient as it yielded enough reflections for the structure solution and refinement.

Finally, the quality of the recorded data depends on the comparability of the intensities recorded on different frames during the tomography. Other than the possibly induced structural changes by the incident beam, a shift in the diffracting crystal volume can introduce difficulties in the merging of the data from different frames. When the incident beam is smaller than the crystal, a shift of the beam with respect to the crystal (or the inverse) can lead to a change in thickness of the diffracting volume, a different orientation if the crystal is bent or a different density of structural defects. In the present case, LD-EDT not only preserves the same amount of crystal volume in the incident beam during the tomography, but even the identical volume, therefore keeping all the characteristics like defect density identical for all frames.

Here we have applied LD-EDT to four very different crystal structures: a complex oxide, a metal-organic framework, a mineral, and an oxide with mixed occupancy on one site, the latter three being very beam sensitive. The LD-EDT method maximizes the signal-to-noise ratio in all diffraction frames, allowing to use the incident electron dose very parsimoniously. The high quality of the obtained data at very low electron doses allowed us not only to solve the structures, but also to use dynamical diffraction theory to refine these structures. In the case of the mineral Bulachite it was possible to refine the thermal parameters of all atoms individually to meaningful values. The values of U_{iso} for the free water molecules were significantly higher than those for the other oxygen ions which could be attributed to a tendency of the free water molecules to leave the crystal under the effect of the vacuum in the TEM and/or the irradiation. Some of the refined U_{iso} were negative, which is not unusual for electron diffraction data even though the underlying reasons why this happens are not clear. They are therefore not very reliable, further progress has to be made in the data treatment of electron diffraction data in order to achieve a reliability comparable to what is obtained from X-ray diffraction data.

The low doses used in this work (down to 0.1 $e^-/Å^2$) are not only sufficient to address any structural study of beam sensitive minerals or metal-organic frameworks, but also open new application fields for this method for example in structural biology or in pharmaceutics.

Author Contributions: Investigation, H.K., S.K., E.Y. and P.B.; Methodology, H.K. and S.K.; Writing—original draft, H.K., S.K.; Writing—review & editing, H.K., S.K., E.Y. and P.B. All authors have read and agreed to the published version of the manuscript.

Funding: This research received no external funding.

Institutional Review Board Statement: Not applicable.

Informed Consent Statement: Not applicable.

Data Availability Statement: Data is contained within the article.

Acknowledgments: The authors are grateful to Dominique Luneau, Celine Darie, Ian Grey, Thomas Mangin, and Lei Ding for providing the samples of this study.

Conflicts of Interest: The authors declare no conflict of interest.

References

1. Henderson, R. The potential and limitations of neutrons, electrons and X-rays for atomic resolution microscopy of unstained biological molecules. *Q. Rev. Biophys.* **1995**, *28*, 171–193. [CrossRef] [PubMed]
2. Kolb, U.; Gorelik, T.; Kubel, C.; Otten, M.T.; Hubert, D. Towards automated diffraction tomography: Part I—Data acquisition. *Ultramicroscopy* **2007**, *107*, 507–513. [CrossRef] [PubMed]
3. Kolb, U.; Gorelik, T.; Otten, M.T. Towards automated diffraction tomography. Part II—Cell parameter determination. *Ultramicroscopy* **2008**, *108*, 763–772. [CrossRef] [PubMed]
4. Mugnaioli, E.; Gorelik, T.; Kolb, U. "Ab initio" structure solution from electron diffraction data obtained by a combination of automated diffraction tomography and precession technique. *Ultramicroscopy* **2009**, *109*, 758–765. [CrossRef] [PubMed]
5. Nannenga, B.; Shi, D.; Leslie, A.; Gonen, T. High-resolution structure determination by continuous rotation data collection in MicroED. *Nat. Methods* **2014**, *11*, 927–930. [CrossRef] [PubMed]
6. Gemmi, M.; la Placa, M.G.I.; Galanis, A.S.; Rauch, E.F.; Nicolopoulos, S. Fast electron diffraction tomography. *J. Appl. Crystallogr.* **2015**, *48*, 718–727. [CrossRef]
7. Gemmi, M.; Mugnaioli, E.; Gorelik, T.E.; Kolb, U.; Palatinus, L.; Boullay, P.; Hovmöller, S.; Abrahams, J.P. 3D Electron Diffraction: The Nanocrystallography Revolution. *ACS Cent. Sci.* **2019**, *5*, 1315–1329. [CrossRef]
8. Palatinus, L.; Corrêa, C.A.; Steciuk, G.; Jacob, D.; Roussel, P.; Boullay, P.; Klementova, M.; Gemmi, M.; Kopecek, J.; Domeneghetti, M.C.; et al. Structure refinement using precession electron diffraction tomography and dynamical diffraction: Tests on experimental data. *Acta Cryst. B* **2015**, *71*, 740–751. [CrossRef]
9. Palatinus, L.; Brázda, P.; Boullay, P.; Perez, O.; Klementová, M.; Petit, S.; Eigner, V.; Zaarour, M.; Mintova, S. Hydrogen positions in single nanocrystals revealed by electron diffraction. *Science* **2017**, *355*, 166–169. [CrossRef]
10. Ge, M.; Zou, X.; Huang, Z. Three-Dimensional Electron Diffraction for Structural Analysis of Beam-Sensitive Metal-Organic Frameworks. *Crystals* **2021**, *11*, 263. [CrossRef]
11. Clabbers, M.; Xu, H. Macromolecular crystallography using microcrystal electron diffraction. *Acta Cryst.* **2021**, *77*, 313–324. [CrossRef] [PubMed]
12. Glaeser, R.M. Specimen Behavior in the Electron Beam. *Methods Enzymol.* **2016**, *579*, 19–50. [PubMed]
13. Kodjikian, S.; Klein, H. Low-dose electron diffraction tomography (LD-EDT). *Ultramicroscopy* **2019**, *200*, 12–19. [CrossRef] [PubMed]
14. Klein, H.; David, J. The quality of precession electron diffraction data is higher than necessary for structure solution of unknown crystalline phases. *Acta Crystallogr.* **2011**, *67*, 297–302. [CrossRef] [PubMed]
15. Vincent, R.; Midgley, P.A. Double conical beam-rocking system for measurement of integrated electron diffraction intensities. *Ultramicroscopy* **1994**, *53*, 271. [CrossRef]
16. Palatinus, L. *PETS—Program for Analysis of Electron Diffraction Data*; Institute of Physics of the AS CR: Prague, Czech Republic, 2011.
17. Burla, M.C.; Caliandro, R.; Carrozzini, B.; Cascarano, G.L.; Cuocci, C.; Giacovazzo, C.; Mallamo, M.; Mazzone, A.; Polidori, G. Crystal structure determination and refinement *viaSIR2014*. *J. Appl. Crystallogr.* **2015**, *48*, 306–309. [CrossRef]
18. Palatinus, L.; Chapuis, G. SUPERFLIP—A computer program for the solution of crystal structures by charge flipping in arbitrary dimensions. *J. Appl. Cryst.* **2007**, *40*, 786–790. [CrossRef]
19. Petříček, V.; Dusek, M.; Palatinus, L. Crystallographic Computing System JANA2006: General features. *Z. Kristallogr.* **2014**, *229*, 345–352. [CrossRef]

20. Klein, H.; Kodjikian, S.; Philippe, R.; Ding, L.; Colin, C.V.; Darie, C.; Bordet, P. Three different Ge environments in a new $Sr_5CuGe_9O_{24}$ phase synthesized at high pressure and high temperature. *Acta Cryst.* **2020**, *B76*, 727–732. [CrossRef]
21. Poulsen, R.D.; Jørgensen, M.R.V.; Overgaard, J.; Larsen, F.K.; Morgenroth, W.; Graber, T.; Chen, Y.-S.; Iversen, B.B. Synchrotron X-ray Charge-Density Study of Coordination Polymer [Mn $(HCOO)_2$ $(H_2O)_2$]. *Chem. Eur. J.* **2007**, *13*, 9775–9790. [CrossRef]
22. Grey, I.E.; Yoruk, E.; Kodjikian, S.; Klein, H.; Bougerol, C.; Brand, H.E.A.; Bordet, P.; Mumme, W.G.; Favreau, G.; Mills, S.J. Bulachite, $[Al6(AsO_4)_3(OH)_9(H_2O)_4]$ $2H_2O$ from Cap Garonne, France: Crystal structure and formation from a higher hydrate. *Mineral. Mag.* **2020**, *84*, 608–615. [CrossRef]
23. Redhammer, G.J.; Roth, G.; Topa, D.; Amthauer, G. Synthetic aenigmatite analog $Na_2(Mn_{5.26}Na_{0.74})Ge_6O_{20}$: Structure and crystal chemical considerations. *Acta Cryst.* **2008**, *C64*, i21–i26. [CrossRef] [PubMed]
24. Prewitt, C.T.; Sleight, A.W. Garnet-Like Structures of High-Pressure Cadmium Germanate and Calcium Germanate. *Science* **1969**, *163*, 386–387. [CrossRef] [PubMed]
25. Rodriguez-Carvajal, J. Recent advances in magnetic structure determination by neutron powder diffraction. *Phys. B* **1993**, *192*, 55–69. [CrossRef]

symmetry

MDPI

Article

Ruddlesden–Popper Faults in $NdNiO_3$ Thin Films

Chao Yang [1,*], Yi Wang [1,2,*], Daniel Putzky [1], Wilfried Sigle [1], Hongguang Wang [1], Roberto A. Ortiz [1], Gennady Logvenov [1], Eva Benckiser [1], Bernhard Keimer [1] and Peter A. van Aken [1]

1 Max Planck Institute for Solid State Research, 70569 Stuttgart, Germany; d.putzky@fkf.mpg.de (D.P.); w.sigle@fkf.mpg.de (W.S.); hgwang@fkf.mpg.de (H.W.); r.ortiz@fkf.mpg.de (R.A.O.); g.logvenov@fkf.mpg.de (G.L.); benckise@fkf.mpg.de (E.B.); b.keimer@fkf.mpg.de (B.K.); p.vanaken@fkf.mpg.de (P.A.v.A.)

2 Center for Microscopy and Analysis, Nanjing University of Aeronautics and Astronautics, Nanjing 210016, China

* Correspondence: c.yang@fkf.mpg.de (C.Y.); wang.yi@nuaa.edu.cn (Y.W.)

Abstract: The $NdNiO_3$ (NNO) system has attracted a considerable amount of attention owing to the discovery of superconductivity in $Nd_{0.8}Sr_{0.2}NiO_2$. In rare-earth nickelates, Ruddlesden–Popper (RP) faults play a significant role in functional properties, motivating our exploration of its microstructural characteristics and the electronic structure. Here, we employed aberration-corrected scanning transmission electron microscopy and spectroscopy to study a $NdNiO_3$ film grown by layer-by-layer molecular beam epitaxy (MBE). We found RP faults with multiple configurations in high-angle annular dark-field images. Elemental intermixing occurs at the $SrTiO_3$–$NdNiO_3$ interface and in the RP fault regions. Quantitative analysis of the variation in lattice constants indicates that large strains exist around the substrate–film interface. We demonstrate that the Ni valence change around RP faults is related to a strain and structure variation. This work provides insights into the microstructure and electronic-structure modifications around RP faults in nickelates.

Keywords: Ruddlesden–Popper faults; NNO thin films; EELS; HAADF image

1. Introduction

The recent discovery of nickel-based superconductors has filled the gap in nickel-based oxide materials in superconducting systems [1–5]. The infinite-layer phase $NdNiO_2$ can only be synthesized from the precursor ABO_3 perovskite structure by removing apical oxygen atoms from NiO_6 octahedra through soft-chemistry topotactical reduction [6]. For stabilization of the superconducting phase, chemical doping with divalent Sr replacing trivalent Nd is crucial. Optimal doping with the highest superconducting transition temperature (around 15 K) has been found in $Nd_{0.8}Sr_{0.2}NiO_2$ thin films [1]. During pulsed-laser deposition growth of Sr-doped thin $NdNiO_3$ films, a strong tendency to form Ruddlesden–Popper (RP) faults has been reported [2]. In addition, the epitaxial strain from the substrate can induce the formation of RP faults [7,8]. These RP faults display an atomic structure, where the inclusion of an additional AO layer breaks the long-range order of the ABO_3 perovskite phase [3,7–17]. As the microstructure of the entire film determines its electric properties, investigation of the defect structure and associated variations in the local electronic structure in nickelates is indispensable. Moreover, RP structures are related to a variety of physical and chemical properties, e.g., electro-catalytic activities [11], microwave dielectric performance [18], magnetic properties [19], and ferroelectric properties [20].

Epitaxial strain and/or a cation non-stoichiometry, e.g., an excess of A or B in the ABO_3 structure, can induce the generation of different RP faults that consist of an intergrowth of rock-salt-type and perovskite-type building blocks [11,12,21,22]. A single rock-salt layer can be described as an $a/2$ <111> or $a/2$ <110> stacking fault displaying a zigzag arrangement of cations [17]. The $(AO)(ABO_3)_n$ RP structure forms when a rock-salt layer separates the n-layered structure of $n(ABO_3)$ [9,23,24]. Detemple et al. [13] reported on three-dimensionally arranged RP faults in $LaNiO_3$–$LaAlO_3$ superlattices, where the authors observed an identical contrast of A and B columns in the ABO_3 structure in high-angle annular dark-field (HAADF) scanning transmission electron microscopy (STEM) investigations. Coll et al. [23] identified multiple configurations of RP faults in a $LaNiO_3$ film based on atomistic modeling, experimental HAADF-STEM investigations, and HAADF-image simulations. Bak et al. [11] demonstrated a strong tetragonal Jahn–Teller distortion of the RP-fault plane in $LaNiO_3$, proving a correlation between the octahedral distortion and the electro-catalytic performance of La-based perovskite oxides. However, to the best of our knowledge, influences of different RP-fault types on the variation in the electronic structure and especially in the Ni valence state have not been reported so far for NNO thin films.

Here, we characterize the local atomic structure of different RP-fault configurations and focus on the electronic properties around RP faults in the NNO film, based on advanced aberration-corrected scanning transmission electron microscopy (STEM) and electron energy-loss spectroscopy (EELS). RP-fault formation starts after a critical thickness of around two unit cells above the substrate–film interface, where different RP-fault types are found to be inhomogeneously distributed in the entire film. We identify an elemental intermixing in various RP-fault regions with different structural configurations. In addition, we demonstrate that the variation in the Ni oxidation state around different RP faults relates to the alteration of the chemical composition, possibly triggered by epitaxial strain.

2. Materials and Methods

The $NdNiO_3$ film was grown on a $SrTiO_3$ (001) single-crystal substrate at 630 °C in an ozone–oxygen atmosphere with a pressure of 1.8×10^{-5} Torr in an atomic layer-by-layer fashion in a custom-made ozone-assisted MBE setup. Then, it was cooled down after the growth in the same atmosphere. The fluxes of the effusion cells were calibrated prior to growth with a quartz crystal microbalance and monitored during growth by in situ reflection high-energy electron diffraction.

The TEM sample was prepared by automated tripod polishing (in an Allied MultiPrep System) until the sample was thinner than ~15 µm, after which Ar-ion milling at liquid N_2 temperature was performed by an ion polishing system (PIPS, Gatan, model 691) [25]. The milling step was performed at a voltage of 0.5 kV to reduce the damage to the surface of the sample. We used a spherical-aberration-corrected microscope (JEOL JEM-ARM200F, JEOL Co., Ltd., Tokyo, Japan) equipped with a DCOR probe corrector (CEOS GmbH, Heidelberg, Germany) and a Gatan GIF Quantum ERS with a Gatan K2 camera at 200 kV to acquire STEM images and EELS spectra. The convergence angle was set to 20.4 mrad and the collection semi-angles for high-angle annular dark-field (HAADF) imaging and annular bright-field (ABF) imaging were set to 70–300 mrad and 10–20 mrad, respectively. The EELS spectra were acquired at a dispersion of 0.5 eV per channel (with an energy resolution of ~1 eV). The principal component analysis (PCA) method [26] was utilized to optimize the signal and to reduce the noise in the EELS maps.

3. Results and Discussion

As the overview HAADF image in Figure 1a shows, the film thickness is around 9 nm, in good agreement with the X-ray reflectometry, and a large number of RP faults are distributed randomly throughout the entire film. The critical thickness for RP-fault formation is around two to three unit cells. In this range, the film grows coherently with the substrate. Similar observations were also made and are discussed in [2]. Above the critical thickness, defects are introduced to accommodate the strain. We note that this growth mode resembles the Stranski–Krastanov mode; however, it differs from it by not forming islands on the coherent NNO layer, but instead forming a continuous film. According to the relationship that the atomic-column brightness is approximately proportional to $Z^{1.7}$ (Z: atomic number), the sufficient contrast difference enables us to distinguish the atomic columns [27]. For example, we can observe a sharp interface between STO and NNO. A high density of irregularly distributed RP faults forms generally three unit cells away from the substrate.

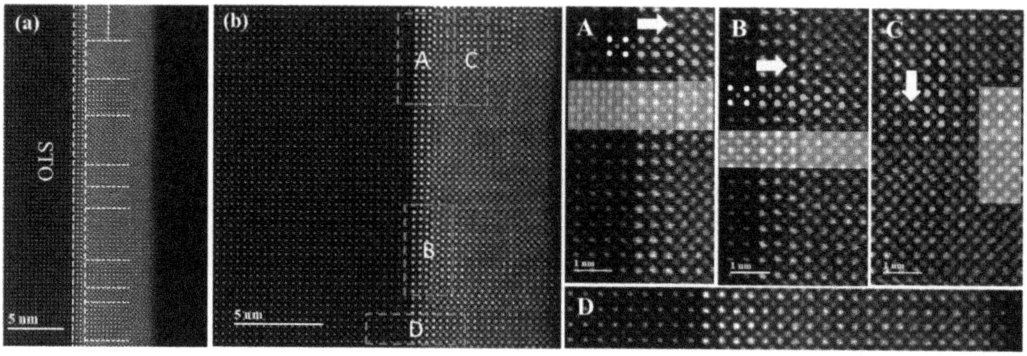

Figure 1. (**a**) An overview HAADF image of a $NdNiO_3$ film grown on a (001) $SrTiO_3$ single-crystal substrate. (**b**) Enlarged image showing different types of RP faults and fault-free regions marked by red dashed boxes in the $NdNiO_3$ layer: (**A**) *a*/2 <111> shift; (**B**) single intergrowth layer of $\{NdNiO_3\}$; (**C**) identical contrast of A/B sites in $NdNiO_3$; (**D**) fault-free region. The contrast-inverted ABF images (shaded regions) show the oxygen columns around the RP faults.

Although we cannot exclude non-stoichiometry during film growth as a reason for RP-fault formation, a major driving force in our system is the reduction in the lattice mismatch between STO (3.905 Å) and NNO (pseudocubic lattice constant: 3.807 Å) as we do not observe any misfit dislocation that could alternatively have reduced the strain. In the enlarged HAADF image (Figure 1b), several configurations of RP faults in the NNO layer are marked with red dashed boxes (A, B, and C, respectively). The apparent *a*/2<110> shift for RP fault A, where *a* is the lattice constant of the pseudocubic unit cell, is likely to be the component of the true displacement vector *a*/2<111> projected along the beam direction [24]. RP fault B shows a single intergrowth layer of $\{NdNiO_3\}$, forming a typical '214' structure of the RP phase (Nd_2NiO_4) by combining a rock-salt-type layer and a perovskite-type layer. In the case of RP fault C, the contrast difference between the A and B sites in NNO disappears due to the overlap of the A and B columns along the viewing direction. [13,23]. Additionally, the red dashed box D marks a fault-free region displaying a perovskite structure of the NNO layer.

Furthermore, we can observe the oxygen columns from the inverted ABF images (shaded in Figure 1) around the edges of these RP faults. The slight elongation of Ni–O bonds along the *z*-axis at the plane of RP fault A is apparent (see also Figure S2b), which can result in a significant splitting of the e_g level of the Ni 3d orbital and a relative variation in the O 2p level based on the Jahn–Teller effect [11]. From a layer-by-layer structural perspective [11], the distortion of oxygen anions toward the fault plane can reduce the strong repulsion between two $[NdO]^+$ layers and achieve an energetically favorable structure. Around RP fault B, the elongation of Ni–O bonds is not clear, accompanying a visible distortion of $[NiO_6]$ octahedra (see Figure S3). The reduced contrast of oxygen columns is possibly due to the existence of oxygen vacancies. Additionally, we observe distinct oxygen columns at RP fault C without apparent changes in the oxygen structure. Owing to the overlapping of elements around the edge of RP fault C, we observe a homogeneous distribution of oxygen.

To investigate the effects of RP faults on the electronic character of the NNO film, we performed STEM-EELS measurements with atomic resolution. Figure 2a,d show the HAADF images of RP faults A and B, respectively. For the sake of improving the signal-to-noise ratio, the O-K and Ni-$L_{2,3}$ edges were extracted from the red dashed boxes with a width of one unit cell shown in Figure 2b,e. From the $NdNiO_3$ reference, we observe a strong pre-peak A at the O K edge, which is associated with the hybridization of O 2p and Ni 3d orbitals, reflecting the covalency of Ni–O bonding and the oxidation state of Ni [28,29]. The pre-peak of the O K edge is weak or even absent around the RP faults, which is related to the changes in the local Ni–O distances due to epitaxial strain and RP faults [28,30]. In addition, the strong Nd–O interaction in the rock-salt layer results in a distinct shoulder of peak C. Close to the STO–NNO interface, the intermixing of Ti into Ni causes the appearance of peak B that arises from the Ti 3d–O 2p hybridized states. Peak B almost disappears in cells 4–7. Peak C corresponds to the O 2p states hybridized with Sr 4d in STO or Nd 5d in NNO. Peak D is due to the O 2p states hybridized with Ni 4 sp and Nd 5p states [31]. The L_3 and L_2 peaks of Ni arise from excitation of electrons from the $2p_{3/2}$ and $2p_{1/2}$ core levels to the unoccupied 3d band, which is used to determine the valence state of the transition metal from the white-line ratio and chemical shift [28]. The slight variations in peak positions and crystal-field splitting observed in Figure 2b,e imply changes in the electronic structure of Ni for both types of RP faults. We quantified the white-line ratio, which is shown in Figure 2c,f. From the DigitalMicrograph database [32], we calculated the Ni^{3+} and Ni^{2+} references, which are expected to correspond to the fault-free $NdNiO_3$ and NiO films. These are shown as black and blue dashed lines in Figure 2c,f, respectively. We observe that the L_3/L_2 ratio increases near RP-fault planes and in the fault-free NNO layers close to the STO substrate, suggesting a decreased valence of Ni. In the fault-free NNO layer (the seventh unit cell) far from the fault planes, the valence of Ni is close to 3+, as expected. The Ni L_3/L_2 intensity ratio in RP fault B is a little higher than that in RP fault A, indicating a lower valence of Ni around RP fault B.

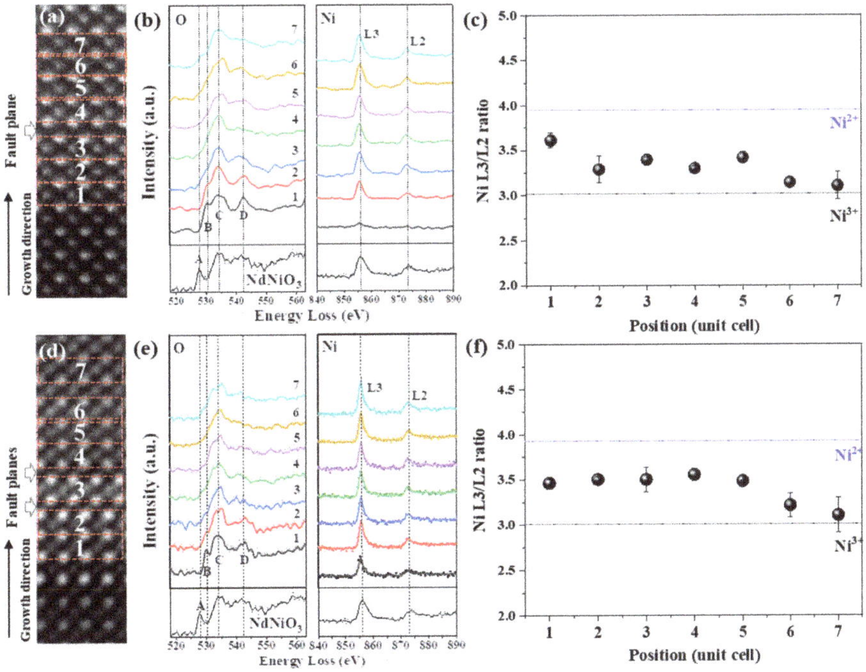

Figure 2. The HAADF images of (**a**) RP fault A and (**d**) RP fault B used for EELS data acquisition. Atomic layers are numbered from 1 to 7 (**b**,**e**) O K and Ni L spectra extracted from the red dashed boxes in the HAADF images of RP faults A and B, respectively. Spectral features are marked by letters A–D (**c**,**f**) The calculated Ni L_3/L_2 white-line intensity ratio around RP fault A and RP fault B, respectively. The Ni^{3+} and Ni^{2+} references are from $NdNiO_3$ and NiO, respectively. The error bar was obtained by measuring different regions.

The atomically resolved EELS maps shown in Figure 3 confirm the atomic arrangement around RP faults A and B. The distribution of cations from the color-coded EELS maps indicates elemental intermixing at the STO–NNO interface. In the RP fault regions, the zigzag arrangement of the Nd atoms and additional intergrowth of an NNO layer verify the atomic structures visible in ADF images. Figure 3a shows the EELS maps of RP fault A, where Ni and Nd columns can be clearly distinguished. In the RP fault B region (Figure 3b), a small amount of Ni is intermixed into Nd columns. Owing to the random distribution of RP faults, it is possible that the observed Nd/Ni signal intermixing arises from the partial overlapping of Ni and Nd. Similarly, the obvious intermixing of Ni and Nd signals at RP fault C shown in Figure S1a is due to the Ni/Nd overlapping, explaining why Ni and Nd columns display the same contrast in ADF images. Moreover, the Ni $L_{2,3}$ and O K edges extracted from fault-free and Ni/Nd overlapping regions show the same shapes and peak positions. The pronounced small pre-peak of the O K edge is consistent with the Ni^{3+} reference in Figure 2b,e.

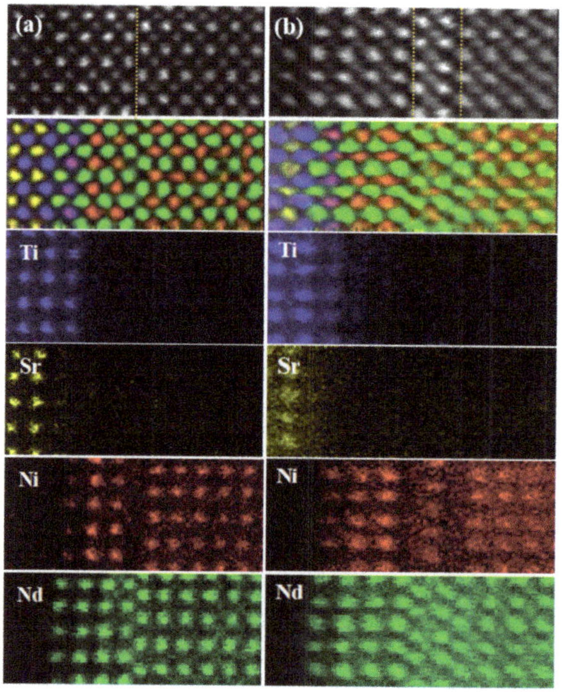

Figure 3. HAADF images of RP fault A (**a**) and RP fault B (**b**). Color-coded elemental maps of Ti (blue), Sr (yellow), Ni (red), and Nd (green) determined from an EELS spectrum image. The RP fault planes are marked with yellow dashed lines. The EELS spectra are shown in Figure S4.

In order to explore the potential mechanism of the valence change in Ni, we quantitatively analyzed the variation in the lattice constant around RP faults A and B. As the lattice constant of STO is a little larger than that of NNO, an in-plane tensile strain of ~2.5% is induced in the NNO film. From Figure 4a,b, the critical thickness of the NNO layer is two unit cells. By calculating the Nd–Nd distance across the STO–NNO interface and across the RP fault plane from the HAADF images, we found that the in-plane tensile strain is maintained at ~2.5% for the first two unit cells before the formation of RP faults. The in-plane Nd–Nd distance decreases to 3.85 ± 0.04 Å at the fault-free layers far from the RP fault plane shown in Figure 4c,d. According to theoretical calculations, the O stoichiometry in NNO is sensitive to the in-plane tensile strain [33]. The formation energy of an oxygen vacancy becomes negative when the in-plane tensile strain is higher than 1.9% [33]. Therefore, the probability for the generation of oxygen vacancies close to the interface is higher than that in fault-free layers where the in-plane strain is significantly released. In addition, oxygen vacancies are more easily generated in the NiO_2 plane as their formation energy in the NiO_2 plane is about 1.7 eV lower than in the NdO plane under a tensile strain [33]. From the inverted ABF image in Figure S2, we can clearly observe that the contrast of oxygen columns in the NiO_2 planes (marked with white arrows) is smaller than that in the NdO planes (marked with orange arrows), indicative of the possible existence of oxygen vacancies in our sample. Thus, the strain-induced oxygen vacancies in the region close to the STO–NNO interface can result in a decrease in the Ni valence. In addition, the sharp drop of the out-of-plane Nd–Nd distance to ~3.5 Å observed a few atomic columns away from the fault plane suggests a large out-of-plane contraction, which gradually increases to ~3.8 Å across the fault-plane and into the fault-free region as shown in Figure 4c,d. Therefore, the generation of RP faults effectively relaxes the epitaxial strain from the substrate, which is consistent with the previous report [12].

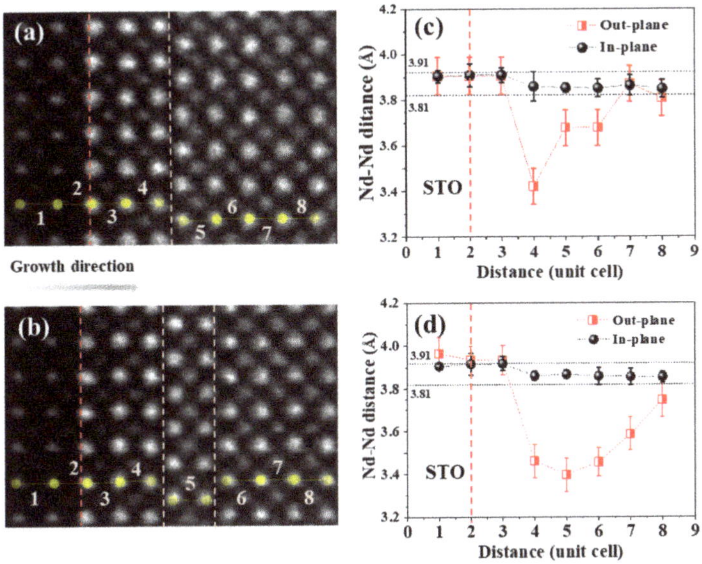

Figure 4. (**a**,**b**) HAADF images of RP fault A and B, respectively, used for the quantitative analysis of Nd–Nd distances. The numbers indicate the positions of unit cells. (**c**,**d**) The Nd–Nd distances along in-plane and out-of-plane directions across RP fault A and B, respectively. The red dashed lines indicate the STO–NNO interface. The orange dashed lines mark the fault planes.

It is worth mentioning that the average oxidation state of nickel varies between 2+ and 3+ for different n in the RP structure ($Nd_{n+1}Ni_nO_{3n+1}$). In principle, Ni is divalent in the '214' end member of the RP structure (Nd_2NiO_4), where the $[NdO]^+$ layer donates an additional electron for the NNO layer [34]. However, the measured Ni valence in RP fault B shows a mixed valence between 2+ and 3+. The RP fault B structure is similar to the Nd_2NiO_4 structure, but the $[NdO]^+$ layer is shared by adjacent NNO layers. We can observe the irregular distribution of different types of RP faults and fault-free NNO layers in Figure 1b. As mentioned above, at the NNO layer close to the substrate, the tensile strain can induce the formation of oxygen vacancies, which means that the additional electron can be partially compensated for by oxygen vacancies in the adjacent NNO layer. Thus, the Ni valence can be higher than 2+ in the intergrowth layer of NNO. Given that the Ni/Nd signal intermixing occurs at the intergrowth layer of NNO, it is possible that the measured

Ni signal partially arises from the fault-free NNO layer, which can also lead to a mixed Ni valence in RP fault B. In terms of RP fault A, there is no obvious elemental intermixing and Ni ions show a higher valence (closer to 3+) than in RP fault B. The variation in the Ni oxidation state is mainly ascribed to the additional NdO layer and the possible oxygen vacancies.

4. Conclusions

Based on STEM and EELS measurements, we systematically studied the atomic structure of RP faults and the effects of RP faults on the local electronic configuration of Ni in a $NdNiO_3$ film. Several RP faults were detected, showing, e.g., $a/2$<111> shifts, intergrowth layers of $NdNiO_3$, and overlapping Nd and Ni columns. The critical thickness for RP fault formation is around two unit cells and no misfit dislocations are visible. The generation of a high density of RP faults effectively relaxes the epitaxial strain from the substrate. We observed the elongation and distortion of the oxygen sublattice around RP-fault planes, which is theoretically beneficial to reduce the strong repulsion between two neighboring $[NdO]^+$ layers. We analyzed the variation in the Ni oxidation state around RP faults in NNO. The changed valence of Ni around RP-fault planes is mainly associated with the variations in the strain and atomic structure. On the one side, the tensile strain might induce oxygen vacancies at the fault-free layer of $NdNiO_3$ close to the substrate. On the other side, the additional $[NdO]^+$ layer can lead to a reduction in the valence of Ni in RP fault regions. This work highlights how the distribution of strain and atomic structure changes around RP faults lead to a spatially varying electronic configuration in $NdNiO_3$.

Supplementary Materials: The following supporting information can be downloaded at: https://www.mdpi.com/article/10.3390/sym14030464/s1. Figure S1. (a) HAADF image of RP fault C and corresponding EELS maps of Ni (red) and Nd (green). (b) Ni L edge and (c) O K edge extracted from regions marked by red dashed boxes in the HAADF image. Figure S2. (a) HAADF and (b) inverted ABF images of RP fault A. The white arrows indicate basal O, and orange arrows represent apical O. Ni columns are marked by red circles. Figure S3. (a) HAADF and (b) inverted ABF image of RP fault B. The yellow circles indicate O columns, and red spots mark the Ni columns. White and green arrows indicate the fault planes. Figure S4. (a) HAADF image used for EELS data acquisition. (b,c) EELS spectra extracted from the red dashed boxes in the HAADF image.

Author Contributions: Methodology, C.Y., D.P., R.A.O. and G.L.; formal analysis, C.Y., H.W.; resources, B.K. and P.A.v.A.; writing—original draft, C.Y.; writing—review and editing, W.S., H.W., E.B. and P.A.v.A.; supervision, Y.W., W.S. and P.A.v.A. All authors have read and agreed to the published version of the manuscript.

Funding: The authors acknowledge the funding support from the European Union's Horizon 2020 research and innovation program under Grant Agreement No. 823717-ESTEEM3 and the Deutsche Forschungsgemeinschaft (DFG, German Research Foundation) TRR 80-project G1.

Institutional Review Board Statement: Not applicable.

Informed Consent Statement: Not applicable.

Data Availability Statement: The datasets generated during and/or analysed during the current study are available from the corresponding author on reasonable request.

Acknowledgments: We are thankful to U. Salzberger for the TEM sample preparation and K. Hahn and P. Kopold for the help with the TEM.

Conflicts of Interest: The authors declare no conflict of interest.

References

1. Li, D.; Lee, K.; Wang, B.Y.; Osada, M.; Crossley, S.; Lee, H.R.; Cui, Y.; Hikita, Y.; Hwang, H.Y. Superconductivity in an infinite-layer nickelate. *Nature* **2019**, *572*, 624–627. [CrossRef] [PubMed]
2. Lee, K.; Goodge, B.H.; Li, D.; Osada, M.; Wang, B.Y.; Cui, Y.; Kourkoutis, L.F.; Hwang, H.Y. Aspects of the synthesis of thin film superconducting infinite-layer nickelates. *APL Mater.* **2020**, *8*, 041107. [CrossRef]

3. Li, D.; Wang, B.Y.; Lee, K.; Harvey, S.P.; Osada, M.; Goodge, B.H.; Kourkoutis, L.F.; Hwang, H.Y. Superconducting Dome in $Nd_{1-x}SrxNiO_2$ Infinite Layer Films. *Phys. Rev. Lett.* **2020**, *125*, 027001. [CrossRef] [PubMed]
4. Hepting, M.; Li, D.; Jia, C.J.; Lu, H.; Paris, E.; Tseng, Y.; Feng, X.; Osada, M.; Been, E.; Hikita, Y.; et al. Electronic structure of the parent compound of superconducting infinite-layer nickelates. *Nat. Mater.* **2020**, *19*, 381–385. [CrossRef] [PubMed]
5. Goodge, B.H.; Li, D.; Lee, K.; Osada, M.; Wang, B.Y.; Sawatzky, G.A.; Hwang, H.Y.; Kourkoutis, L.F. Doping evolution of the Mott–Hubbard landscape in infinite-layer nickelates. *Proc. Natl. Acad. Sci. USA* **2021**, *118*, e2007683118. [CrossRef] [PubMed]
6. Hayward, M.A.; Green, M.A.; Rosseinsky, M.J.; Sloan, J. Sodium Hydride as a Powerful Reducing Agent for Topotactic Oxide Deintercalation: Synthesis and Characterization of the Nickel(I) Oxide $LaNiO_2$. *J. Am. Chem. Soc.* **1999**, *121*, 8843–8854. [CrossRef]
7. Bak, J.; Chung, S.-Y. Observation of fault-free coherent layer during Ruddlesden-Popper faults generation in $LaNiO_3$ thin films. *J. Korean Ceram. Soc.* **2020**, *58*, 169–177. [CrossRef]
8. Qi, H.; Chen, X.; Wu, M.; Benckise, E.; Logvenov, G.; Christiani, G.; Keimer, B.; Kaiser, U. Formation mechanism of Ruddlesden-Popper faults in compressive-strained ABO_3 perovskite superlattices. *Nanoscale* **2021**, *13*, 20663–20669. [CrossRef]
9. Ruddlesden, S.N.; Popper, P. New compounds of the K_2NIF_4 type. *Acta Crystallogr.* **1957**, *10*, 538–539. [CrossRef]
10. Yu, Y.; Zhang, D.; Yang, P. Ruddlesden-Popper Phase in Two-Dimensional Inorganic Halide Perovskites: A Plausible Model and the Supporting Observations. *Nano Lett.* **2017**, *17*, 5489–5494. [CrossRef]
11. Bak, J.; Bin Bae, H.; Kim, J.; Oh, J.; Chung, S.-Y. Formation of Two-Dimensional Homologous Faults and Oxygen Electrocatalytic Activities in a Perovskite Nickelate. *Nano Lett.* **2017**, *17*, 3126–3132. [CrossRef] [PubMed]
12. Bak, J.; Bin Bae, H.; Oh, C.; Son, J.; Chung, S.-Y. Effect of Lattice Strain on the Formation of Ruddlesden-Popper Faults in Heteroepitaxial $LaNiO_3$ for Oxygen Evolution Electrocatalysis. *J. Phys. Chem. Lett.* **2020**, *11*, 7253–7260. [CrossRef] [PubMed]
13. Detemple, E.; Ramasse, Q.M.; Sigle, W.; Cristiani, G.; Habermeier, H.U.; Keimer, B.; Van Aken, P.A. Ruddlesden-Popper faults in $LaNiO_3/LaAlO_3$ superlattices. *J. Appl. Phys.* **2012**, *112*, 013509. [CrossRef]
14. Thind, A.S.; Luo, G.; Hachtel, J.A.; Morrell, M.V.; Cho, S.B.; Borisevich, A.Y.; Idrobo, J.; Xing, Y.; Mishra, R. Atomic Structure and Electrical Activity of Grain Boundaries and Ruddlesden-Popper Faults in Cesium Lead Bromide Perovskite. *Adv. Mater.* **2019**, *31*, e1805047. [CrossRef] [PubMed]
15. Li, Z.; Guo, W.; Zhang, T.T.; Song, J.H.; Gao, T.Y.; Gu, Z.B.; Nie, Y.F. Epitaxial growth and electronic structure of Ruddlesden-Popper nickelates (Lan+1NinO3n+1, n = 1–5). *APL Mater.* **2020**, *8*, 091112. [CrossRef]
16. Wang, W.Y.; Tang, Y.L.; Zhu, Y.L.; Suriyaprakash, J.; Xu, Y.-B.; Liu, Y.; Gao, B.; Cheong, S.-W.; Ma, X.L. Atomic mapping of Ruddlesden-Popper faults in transparent conducting BaSnO3-based thin films. *Sci. Rep.* **2015**, *5*, 16097. [CrossRef]
17. Gauquelin, N.; Benckiser, E.; Kinyanjui, M.K.; Wu, M.; Lu, Y.; Christiani, G.; Logvenov, G.; Habermeier, H.U.; Kaiser, U.; Keimer, B.; et al. Atomically resolved EELS mapping of the interfacial structure of epitaxially strained $LaNiO_3/LaAlO_3$ superlattices. *Phys. Rev. B* **2014**, *90*, 195140. [CrossRef]
18. Lee, C.-H.; Orloff, N.; Birol, T.; Zhu, Y.; Goian, V.; Rocas, E.; Haislmaier, R.; Vlahos, E.; Mundy, J.A.; Kourkoutis, L.; et al. Exploiting dimensionality and defect mitigation to create tunable microwave dielectrics. *Nature* **2013**, *502*, 532–536. [CrossRef]
19. Jing, H.M.; Cheng, S.; Mi, S.B.; Lu, L.; Liu, M.; Cheng, S.D.; Jia, C.L. Formation of Ruddlesden-Popper faults and their effect on the magnetic properties in $Pr_{0.5}Sr_{0.5}CoO_3$ thin films. *ACS Appl. Mater. Interfaces* **2018**, *10*, 1428–1433. [CrossRef]
20. Oh, Y.S.; Luo, X.; Huang, F.-T.; Wang, Y.; Cheong, S.-W. Experimental demonstration of hybrid improper ferroelectricity and the presence of abundant charged walls in $(Ca, Sr)_3Ti_2O_7$ crystals. *Nat. Mater.* **2015**, *14*, 407–413. [CrossRef]
21. Wang, W.; Zhang, H.; Shen, X.; Guan, X.; Yao, Y.; Wang, Y.; Sun, J.; Yu, R. Atomic structures of Ruddlesden-Popper faults in $LaCoO_3/SrRuO_3$ multilayer thin films induced by epitaxial strain. *J. Cryst. Growth* **2018**, *490*, 110–115. [CrossRef]
22. Mundet, B.; Jareño, J.; Gazquez, J.; Varela, M.; Obradors, X.; Puig, T. Defect landscape and electrical properties in solution-derived $LaNiO_3$ and $NdNiO_3$ epitaxial thin films. *Phys. Rev. Mater.* **2018**, *2*, 063607. [CrossRef]
23. Coll, C.; López-Conesa, L.; Rebled, J.M.; Magén, C.; Sánchez, F.; Fontcuberta, J.; Estradé, S.; Peiró, F. Simulation of STEM-HAADF Image Contrast of Ruddlesden-Popper Faulted $LaNiO_3$ Thin Films. *J. Phys. Chem. C* **2017**, *121*, 9300–9304. [CrossRef]
24. Palgrave, R.G.; Borisov, P.; Dyer, M.S.; McMitchell, S.R.C.; Darling, G.R.; Claridge, J.B.; Batuk, M.; Tan, H.; Tian, H.; Verbeeck, J.; et al. Artificial Construction of the Layered Ruddlesden-Popper Manganite $La_2Sr_2Mn_3O_{10}$ by Reflection High Energy Electron Diffraction Monitored Pulsed Laser Deposition. *J. Am. Chem. Soc.* **2012**, *134*, 7700–7714. [CrossRef] [PubMed]
25. Wang, H.; Srot, V.; Fenk, B.; Laskin, G.; Mannhart, J.; van Aken, P.A. An optimized TEM specimen preparation method of quantum nanostructures. *Micron* **2021**, *140*, 102979. [CrossRef]
26. Lucas, G.; Burdet, P.; Cantoni, M.; Hébert, C. Multivariate statistical analysis as a tool for the segmentation of 3D spectral data. *Micron* **2013**, *52-53*, 49–56. [CrossRef]
27. Nellist, P.; Pennycook, S. Incoherent imaging using dynamically scattered coherent electrons. *Ultramicroscopy* **1999**, *78*, 111–124. [CrossRef]
28. Tan, H.; Verbeeck, J.; Abakumov, A.; Van Tendeloo, G. Oxidation state and chemical shift investigation in transition metal oxides by EELS. *Ultramicroscopy* **2012**, *116*, 24–33. [CrossRef]
29. Ahn, C.C. *Transmission Electron Energy-Loss Spectrometry in Materials Science and the EELS Atlas*, 2nd ed.; Wiley: Darmstadt, Germany, 2004.
30. Phillips, P.J.; Rui, X.; Georgescu, A.B.; Disa, A.; Longo, P.; Okunishi, E.; Walker, F.; Ahn, C.H.; Ismail-Beigi, S.; Klie, R. Experimental verification of orbital engineering at the atomic scale: Charge transfer and symmetry breaking in nickelate heterostructures. *Phys. Rev. B* **2017**, *95*, 205131. [CrossRef]

31. Mundet, B.; Domínguez, C.; Fowlie, J.; Gibert, M.; Triscone, J.-M.; Alexander, D.T.L. Near-Atomic-Scale Mapping of Electronic Phases in Rare Earth Nickelate Superlattices. *Nano Lett.* **2021**, *21*, 2436–2443. [CrossRef]
32. Ahn, C.C.; Krivanek, O.L. EELS Atlas. Gatan Inc. & ASU HREM Facility, 1983. Available online: https://eels.info/atlas (accessed on 1 February 2022).
33. Kim, T.H.; Paudel, T.R.; Green, R.J.; Song, K.; Lee, H.-S.; Choi, S.-Y.; Irwin, J.; Noesges, B.; Brillson, L.J.; Rzchowski, M.S.; et al. Strain-driven disproportionation at a correlated oxide metal-insulator transition. *Phys. Rev. B* **2020**, *101*, 121105. [CrossRef]
34. Wrobel, F.; Geisler, B.; Wang, Y.; Christiani, G.; Logvenov, G.; Bluschke, M.; Schierle, E.; van Aken, P.A.; Keimer, B.; Pentcheva, R.; et al. Digital modulation of the nickel valence state in a cuprate-nickelate heterostructure. *Phys. Rev. Mater.* **2018**, *2*, 035001. [CrossRef]

symmetry

MDPI

Article

Comprehensive Study of Li^{+}/Ni^{2+} Disorder in Ni-Rich NMCs Cathodes for Li-Ion Batteries

Elena D. Orlova [1], Aleksandra A. Savina [1,*], Sergey A. Abakumov [2], Anatolii V. Morozov [1] and Artem M. Abakumov [1]

[1] Center for Energy Science and Technology, Skolkovo Institute of Science and Technology, Nobel Str. 3, 121205 Moscow, Russia; Elena.Orlova2@skoltech.ru (E.D.O.); Anatolii.Morozov@skoltech.ru (A.V.M.); A.Abakumov@skoltech.ru (A.M.A.)

[2] Laboratory for Molecular Imaging and Photonics, KU Leuven, Celestijnenlaan 200F, B-3001 Leuven, Belgium; sergey.abakumov@kuleuven.be

* Correspondence: A.Savina@skoltech.ru; Tel.: +79-246571299

Abstract: The layered oxides $LiNi_xMn_yCo_zO_2$ (NMCs, x + y + z = 1) with high nickel content ($x \geq 0.6$, Ni-rich NMCs) are promising high-energy density-positive electrode materials for Li-ion batteries. Their electrochemical properties depend on Li^{+}/Ni^{2+} cation disordering originating from the proximity of the Li^{+} and Ni^{2+} ionic radii. We synthesized a series of the $LiNi_{0.8}Mn_{0.1}Co_{0.1}O_2$ NMC811 adopting two different disordering schemes: Ni for Li substitution at the Li site in the samples finally annealed in air, and close to Ni↔Li antisite disorder in the oxygen-annealed samples. The defect formation scenario was revealed with Rietveld refinement from powder X-ray diffraction data, and then the reliability of semi-quantitative parameters, such as I_{003}/I_{104} integral intensity ratio and $c/(2\sqrt{6}a)$ ratio of pseudocubic subcell parameters, was verified against the refined defect concentrations. The I_{003}/I_{104} ratio can serve as a quantitative measure of $g(Ni_{Li})$ only after explicit correction of intensities for preferred orientation. Being normalized by the total scattering power of the unit cell, the I_{003}/I_{104} ratio depends linearly on $g(Ni_{Li})$ for each disordering scheme. The $c/(2\sqrt{6}a)$ ratio appears to be not reliable and cannot be used for a quantitative estimate of $g(Ni_{Li})$. In turn, the volume of the $R\bar{3}m$ unit cell correlates linearly with $g(Ni_{Li})$, at least for defect concentrations not exceeding 5%. The microscopy techniques such as high-resolution high-angle annular dark-field scanning transmission electron microscopy (HAADF-STEM) and electron diffraction tomography (EDT) allow us to study the materials locally, still, there is no proper quantitative approach for comprehensive analysis of defects. In the present work, the TEM-assisted quantitative Li^{+}/Ni^{2+} disordering analysis with EDT and HAADF-STEM in six Ni-rich NMC samples with various defects content is demonstrated. Noteworthy, while PXRD and EDT methods demonstrate overall defect amounts, HAADF-STEM allows us to quantitatively distinguish regions with various disordering extents. Therefore, the combination of mentioned PXRD and TEM methods gives the full picture of Li^{+}/Ni^{2+} mixing defects in Ni-rich NMCs.

Keywords: Li-ion battery; cathode; layered oxide; Ni-rich NMCs; Li^{+}/Ni^{2+} anti-site defects; transmission electron microscopy; electron diffraction tomography; qualitative high-resolution HAADF-STEM

1. Introduction

Owing to higher specific capacity, higher energy density and reduced use of high-cost and relatively scarce cobalt compared to conventional $LiCoO_2$ positive electrode (cathode) for Li-ion batteries (LIBs), layered mixed lithium and transition metals (TM) oxides $LiNi_xMn_yCo_zO_2$ (NMCs, x + y + z = 1) are considered to be the most promising candidates for positive electrodes (cathodes) for next-generation LIBs [1,2]. NMCs with high nickel content (termed Ni-rich NMCs, where $x \geq 0.6$) are of particularly great interest, as such materials can provide high specific capacity (up to 220 mAh/g in the

2.7–4.3 V vs. Li/Li^+ potential window), and the energy density of such materials can reach 800 Wh/kg [3].

The crystal structure of NMCs is of α-$NaFeO_2$ type (space group $R\bar{3}m$) consisting of TMO_2 layers built up with edge-sharing TMO_6 octahedra and separated by layers of octahedrally coordinated Li cations, being an ordered derivative of the rock-salt structure. In an ideal case, the Li and TM cations are located in the octahedral Wyckoff *3a* and *3b* sites, respectively, thus occupying the octahedral voids in the underlying "cubic" ("*ccc*" or ABCABC) close packing of oxygen atoms along the [001] direction (also denoted as the O3 structure highlighting the octahedral coordination of Li and presence of three Li layers per repeat period along the *c*-axis [4]). This layered structure provides a two-dimensional space for the transport of lithium ions while the material undergoes electrochemical Li (de)intercalation.

In the Ni-rich NMC compounds, the Ni cations are in the mixed-valent Ni^{2+}-Ni^{3+} state. It is generally known that there is strong tendency of Ni^{2+} to occupy the Li^+ *3b* sites because of similar ionic radii of Ni^{2+} and Li^+ in contrast to larger difference with Ni^{3+} ($r(Ni^{3+})$ = 0.56 Å, $r(Ni^{2+})$ = 0.69 Å, $r(Li^+)$ = 0.76 Å for CN = 6 [5]) and lower energy barrier of Ni^{2+} migration from their native *3b* to Li^+ *3a* sites, promoting Ni^{2+}/Li^+ exchange, compared to lower probability of such exchange for Co^{3+} and Mn^{4+} [6]. The point defects due to the Ni^{2+}/Li^+ exchange are widely held in the literature on NMC cathode materials as anti-site defects, or cation disordering. The Ni^{2+}/Li^+ mixing degree is found to increase with increasing the nickel content in NMCs [7,8]. In general, the Ni^{2+}/Li^+ anti-site defects are believed to negatively affect the electrochemical properties of Ni-rich NMCs. The disordered phases are known to have a smaller distance between the TM layers and, therefore, a higher Li^+ migration energy barrier compared to the well-ordered phase, leading to a drop of Li diffusivity by blocking Li^+ migration paths [9–11].

The Ni^{2+}/Li^+ anti-site defects are not the sole reason for the appearance of the Ni cations at the Li site and/or Li cations at the TM site. The pure anti-site disorder can be described using Kröger–Vink notations as

$$\mathrm{Ni}_{\mathrm{Ni}}^{\times} + \mathrm{Li}_{\mathrm{Li}}^{\times} \rightarrow \mathrm{Li}_{\mathrm{Ni}}^{\prime} + \mathrm{Ni}_{\mathrm{Li}}^{\bullet} \tag{1}$$

This exchange preserves the chemical composition, but in fact, the Li/TM ratio in Ni-rich NMCs is not constant and depends on partial oxygen pressure and temperature during the last stage of the solid-state synthesis [12]:

$$\mathrm{Ni}_{\mathrm{Ni}}^{\times} + 2\mathrm{Li}_{(s)} + \mathrm{O}_2 \rightarrow \mathrm{Li}_{\mathrm{Ni}}^{\prime} + \mathrm{Ni}_{\mathrm{Ni}}^{\bullet} + \mathrm{Li}_{\mathrm{Li}}^{\times} + 2\mathrm{O}_{\mathrm{O}}^{\times} \tag{2}$$

Here the oxygen molecule creates two cationic *3a* and *3b* sites which are both populated with the Li cations ($\mathrm{Li}_{\mathrm{Li}}^{\times} + \mathrm{Li}_{\mathrm{Ni}}^{\prime}$) from the Li source ($\mathrm{Li}_{(s)}$) whereas the charge compensation is achieved by oxidation of the Ni^{2+} to Ni^{3+} cation at its native site ($\mathrm{Ni}_{\mathrm{Ni}}^{\times} \rightarrow \mathrm{Ni}_{\mathrm{Ni}}^{\bullet}$). In the electrochemically delithiated material partial loss of oxygen is possible accompanied by migration of the Ni cations to the Li sites that is expressed with a different equation:

$$V_{\mathrm{Li}}^{\prime} + \mathrm{Ni}_{\mathrm{Ni}}^{\bullet} + \mathrm{O}_{\mathrm{O}}^{\times} \rightarrow \mathrm{Ni}_{\mathrm{Li}}^{\bullet} + V_{\mathrm{Ni}}^{\prime\prime\prime} + V_{\mathrm{O}}^{\bullet\bullet} + 1/2\mathrm{O}_2 \tag{3}$$

Here the Li vacancy is created by electrochemical deintercalation resulting in adding a single positive charge to the Ni site. The resulting Ni and O vacancies annihilate causing gradual transformation of a well-ordered layered structure into a spinel-like structure, and ultimately into a rock-salt type structure. This structure "densification" leads to loss of the electrochemical capacity and may increase flammability hazard via evolved oxygen reacting with organic components of electrolyte [13–16]. Moreover, the different structural organization of the initial (layered) and the resulting disordered structures leads to a strong microstrain in the primary particles during multiple charge/discharge cycles and, as a result, a formation of microcracks in the secondary particles, that in turn results in battery failure due to loss of electronic conduction [10,17,18]. On the contrary, a number of studies, including theoretical ones, showed the positive influence of small

Ni^{2+}/Li^{+} mixing, demonstrating that certain concentrations of anti-site defects in the volume of the sample can benefit the thermal and structural stability of Ni-rich NMCs during electrochemical cycling at high states of charge [7,18–20]. Thus, the exact role of Ni^{2+}/Li^{+} exchange in the cathode material's properties is still under debate.

There is no doubt that whatever the origin of Ni^{2+} at the Li site is (i.e., Equations, (1); or, (2)), these defects appear primarily at the surface [15,21–23]. This is not surprising as the surface might be enriched with Ni^{2+} comparing to the bulk as a consequence of partial reduction due to insufficiently high partial oxygen pressure during the synthesis or due to reductive interactions with the electrolyte components. Thus, apart from the total content of Ni^{2+} at the Li site, its local quantification at the crystal surface is of particular interest as it can be directly related to the synthesis conditions and electrochemical behavior of Ni-rich NMCs. However, to our knowledge, systematic investigations of quantitative distribution of cationic defects from bulk to the surface of the NMC particles have not been reported yet.

Powder X-ray diffraction (PXRD) is the most common technique used for the assessment of the Ni^{2+}/Li^{+} disorder, both quantitatively and semi-quantitatively. The ratio of the integrated intensities of the 003 and 104 reflections (I_{003}/I_{104}) and the $c/(2\sqrt{6}a)$ ratio (indicating the trigonal distortion of the *f.c.c.* cubic subcell) are the two most extensively used semi-quantitative parameters, while the fractional occupancies of Li and Ni ions at *3b* and *3a* sites, respectively, directly determined via Rietveld refinement, are considered to be reliable quantitative parameters. Still, PXRD has limitations such as low sensitivity towards such "light" atoms as Li. Neutron powder diffraction (NPD) is a more powerful technique for quantification of the Ni^{2+}/Li^{+} disordering, particularly in a joint Rietveld refinement with synchrotron X-ray powder diffraction data [24–26]. However, NPD usually requires a substantial amount of the material, which is not easily available from the electrochemical cell where the quantity of the cathode material is limited to dozens of mg. Moreover, both PXRD and NPD are essentially bulk methods, which cannot differentiate the effects originating from the surface and the bulk, inhomogeneous composition, different particle size and shape, etc. Transmission electron microscopy (TEM) is an alternative approach, very suitable for local area analysis [27]. High-resolution high angle annular dark-field scanning transmission electron microscopy (HAADF-STEM) and annular bright-field STEM (ABF-STEM) imaging techniques are widely used to obtain the local structural information, but in most cases in a rather empirical and less quantitative manner [15,28–31]. Nevertheless, the tools based on statistical parameter estimation approaches for retrieving local crystallographic information from the TEM images are already available and can be potentially applied to measuring the cationic disorder in Ni-rich NMCs at an atomic scale [32,33].

Another quantitative technique that was successfully applied to anti-site defects evaluation is electron diffraction tomography (EDT) or 3D-electron diffraction (3D-ED) [34]. Electron diffraction is more sensitive to the "light" atoms than PXRD [35], for example, in NMCs the "visibility" of Li cations with electrons is around three times higher, compared to X-rays [36]. Indeed, EDT has already been used successfully for the quantification of Li and TM exchange in polyanion Li-ion battery cathodes [37]. EDT reduces the locality of structural characterization down to the crystals not larger than 200–500 nm closing the gap between bulk diffraction techniques and TEM images [38].

Thus, comprehensive consideration of various approaches to proper quantification of Ni^{2+}/Li^{+} disorder is required for better understanding and control over the crystal structure of Ni-rich NMCs and, therefore, their electrochemical properties. In the present work, we tried to utilize the capabilities of TEM imaging and diffraction techniques for quantification of the occupancy factors of the Ni and Li sites in a typical Ni-rich NMC material $LiNi_{0.8}Mn_{0.1}Co_{0.1}O_2$ (NMC811) and provide the comparison of the obtained quantitative results with those extracted from various quantitative and semi-quantitative treatments of laboratory powder X-ray diffraction patterns. EDT is demonstrated to be a reliable technique for measuring defect concentration at the level of a single crystallite. We suggest the quantitative analysis of high-resolution HAADF-STEM images via statistical parameter estimation theory as a tool for discriminating the local regions with different degrees of anti-site disorder.

2. Materials and Methods

Six NMC811 samples presented in Table 1 with various degrees of Ni^{2+}/Li^{+} disorder were obtained by co-precipitation of either hydroxide or carbonate transition metal (TM) precursors from sulfates and acetates as transition metals sources followed by annealing of the precursors with $LiOH{\cdot}H_2O$ in either air or oxygen atmosphere. Typically, 2M TM sulfate solution ($NiSO_4{\cdot}6H_2O$, $CoSO_4{\cdot}7H_2O$ and $MnSO_4{\cdot}H_2O$, Sigma Aldrich, St. Louis, MO, USA, ≥99%) or 0.5M TM acetate solution ($(CH_3COO)_2Ni{\cdot}4H_2O$, $(CH_3COO)_2Mn{\cdot}4H_2O$, and $(CH_3COO)_2Co{\cdot}4H_2O$, Sigma Aldrich, St. Louis, MO, USA, ≥99%) with the Ni:Mn:Co = 0.8:0.1:0.1 molar ratio, 4M (for sulfates) and 1M (for acetates) NaOH or 2M (for sulfates) and 0.5M (for acetates) Na_2CO_3 with the appropriate amount of $NH_3{\cdot}H_2O$ were used (Table 1). The sulfate or acetate solution and sodium carbonate or hydroxide were separately pumped into a continuously stirred batch reactor under Ar atmosphere. The pH and temperature of the mixed solution were maintained at 11–11.5 and 50 °C for the hydroxide precipitation and 7.8 and 56 °C for the carbonate one. The resulting precipitates were filtered and washed with deionized water several times to remove residual ions. The washed precursors were dried in a vacuum oven at 90 °C for 12 h. The obtained TM hydroxides or carbonates were mixed with $LiOH{\cdot}H_2O$ with a molar ratio of 1:1.06 or 1:1.03, respectively, and annealed for 12 h at 750 °C in oxygen flow or for 5 h at 500 °C and then for 12 h at 850 °C in air.

Table 1. Synthesis conditions of the NMC811 samples.

Sample	Co-Precipitation Route	TM Source	C ($NH_3{\cdot}H_2O$), mol/L	pH	Annealing Atmosphere	Li/TM Ratio
S1air	hydroxide	sulfates	0.25	11.0	air	1.06
S1ox	hydroxide	sulfates	0.25	11.0	oxygen	1.03
S2air	hydroxide	acetates	0.50	11.5	air	1.06
S2ox	hydroxide	acetates	0.50	11.5	oxygen	1.03
S3air	carbonate	sulfates	0.25	7.8	air	1.06
S3ox	hydroxide	sulfates	0.50	11.5	oxygen	1.03

Powder X-ray diffraction (PXRD) patterns were collected with a Huber G670 Guinier diffractometer with Co-$K_{\alpha 1}$ radiation (λ = 1.78892Å, curved Ge (111) monochromator, transmission geometry image plate detector) at room temperature in the 4–100° 2θ range with the angular step of 0.005°. Rietveld refinement was carried out using the JANA2006 program package [39].

TEM samples were prepared in air by crushing the crystals with an agate mortar and pestle in ethanol and depositing drops of suspension onto a carbon film supported by a copper grid. High angle annular dark-field scanning transmission electron microscopy (HAADF-STEM) images were acquired on a probe aberration-corrected Thermo Fisher Titan Themis Z electron microscope at 200 kV (the probe convergence semiangle is 30 mrad, the inner and outer acceptance angles of the HAADF detector are 57 and 200 mrad, respectively). The quantitative analysis of Ni^{2+}/Li^{+} mixing from high-resolution HAADF-STEM images was performed with the StatSTEM software [33].

Electron diffraction tomography (EDT) data were collected using Thermo Fisher Tecnai G2 (Eindhoven, The Netherlands) microscope at 200 kV by measuring series of selected area electron diffraction patterns taken with an angular interval of 1° in the highest achievable for each crystal angular range. The exact ranges for corresponding crystals are provided in Table S1 of Supporting Information. No electron beam precession was used. Acquisition was started from a random crystal orientation far from the low-index zone axes to minimize multiple scattering effects. The obtained series of diffraction patterns were treated using PETS and JANA 2006 software [33,40] for the reflection search, cluster analysis, indexing, refinement of the unit cell parameters and orientation matrix and final integration of the reflection intensities. The obtained set of intensities and *hkl* values was used for the kinematical structure refinement that is assumed to be reasonable approximation as the materials do not contain atomic entities with high scattering power [36].

3. Results and Discussion

The crystallographic parameters of the NMC811 materials were first obtained through the Rietveld refinement from PXRD data. The PXRD patterns of all six samples (Figure 1) demonstrate well-crystallized single-phase compounds with the layered α-$NaFeO_2$-type structure. The patterns were completely indexed with the hexagonal cell and $R\bar{3}m$ space group with the unit cell parameters and volumes listed in Table 2. The crystal structure refinement was conducted using the following scheme. The *3a* (0,0,0) site was jointly populated by Li and Ni, whereas Li, Ni, Mn and Co were placed at the *3b* (0,0,1/2) site. The occupancy factors g_{Li} and g_{Ni} at the *3a* site were refined with the $g_{Li} + g_{Ni} = 1$ constrain. The occupancy factors g_{Mn} and g_{Co} at the *3b* site were fixed to 0.1 each according to the NMC811 chemical composition, and g_{Li} and g_{Ni} were refined with $g_{Li} + g_{Ni} = 0.8$. For the S1air, S2air and S3air samples the refined g_{Li} at the *3b* site was equal to zero within one standard deviation. Thus, it was further assumed that Li does not populate this crystallographic position. Common atomic displacement parameter (ADP) was refined for all cationic and oxygen sites. Finally, preferred orientation along the [001] direction was taken into account with the March–Dollase function. The resulting low-reliability factors and good agreement between the experimental and calculated PXRD profiles (Figures S1–S6) indicate high reliability of the Rietveld refinements. The crystallographic data, after the Rietveld refinement and parameters of the Ni^{2+}/Li^{+} mixing, are shown in Table 2.

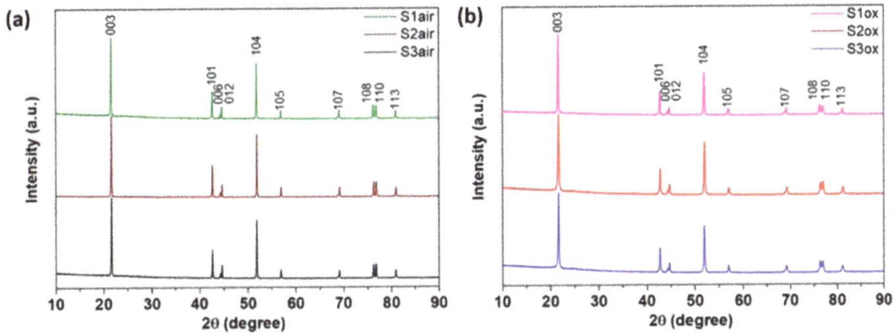

Figure 1. PXRD patterns of the NMC811 (**a**) air-annealed and (**b**) oxygen-annealed samples.

The refined structures and chemical compositions are in general agreement with the sample preparation conditions. Annealing of the samples at higher temperature and lower partial oxygen pressure (850 °C, air, $p(O_2) = 0.21$ atm) is favorable for a lower oxidation state of Ni that results in introducing Ni into the Li site similar to the Equation (3). Indeed, all air-prepared samples demonstrate certain Ni for Li substitution at the *3a* site and full TM occupation of the *3b* site. Larger unit cell volume for the air-annealed samples ($\langle V \rangle = 102.01$ Å^3) compared to that of the oxygen-annealed ones ($\langle V \rangle = 101.46$ Å^3) indicates more reduced Ni. Annealing at lower temperature and higher partial oxygen pressure (750 °C, air, $p(O_2) = 1$ atm) introduces Li into the TM site according to Equation (2) that raises the Ni oxidation state. It should be noted that in neither of the sample the defect concentrations correspond to the pure anti-site scenario as defined with Equation (1).

Table 2. Unit cell parameters, refined atomic coordinates, ADPs, March–Dollase preferred orientation parameter (τ), interatomic distances, parameters of the Ni/Li mixing at the 3*a* and 3*b* sites and reliability factors for the NMC811 materials after Rietveld refinement.

Sample	Refined Composition	*a*, Å	*c*, Å	V, Å^3	z_O	U_{iso}, Å^2	τ	d(Li-O), Å	d(TM-O), Å	Ni in 3*a* Site, %	Li in 3*b* Site, %	R_F, %	R_p, %
S1air	$Li_{0.965}Ni_{0.835}Mn_{0.1}Co_{0.1}O_2$	2.87712(1)	14.2274(1)	101.994(1)	0.24047(8)	0.0049(2)	0.9724(9)	2.1225(7)	1.9651(6)	3.50(7)	0	2.21	1.66
S2air	$Li_{0.975}Ni_{0.825}Mn_{0.1}Co_{0.1}O_2$	2.876568(8)	14.21556(7)	101.8694(7)	0.24095(7)	0.0051(2)	0.9888(7)	2.1173(6)	1.9681(5)	2.45(6)	0	1.24	2.07
S3air	$Li_{0.950}Ni_{0.850}Mn_{0.1}Co_{0.1}O_2$	2.87927(1)	14.2290(1)	102.1578(9)	0.24076(8)	0.0057(2)	0.9823(9)	2.1210(7)	1.9685(6)	5.02(7)	0	2.30	2.00
S1ox	$Li_{0.995}Ni_{0.805}Mn_{0.1}Co_{0.1}O_2$	2.87401(2)	14.2097(2)	101.646(2)	0.24092(7)	0.0049(2)	0.9781(8)	2.1160(7)	1.9664(6)	2.83(6)	2.4(2)	2.00	1.75
S2ox	$Li_{1.013}Ni_{0.787}Mn_{0.1}Co_{0.1}O_2$	2.86994(2)	14.1975(2)	101.271(2)	0.24020(9)	0.0047(3)	0.9804(9)	2.1199(8)	1.9584(7)	2.24(7)	3.5(2)	2.42	1.65
S3ox	$Li_{0.993}Ni_{0.807}Mn_{0.1}Co_{0.1}O_2$	2.87202(3)	14.2040(3)	101.465(3)	0.24062(8)	0.0048(2)	0.9756(9)	2.1175(7)	1.9629(6)	2.61(6)	1.9(2)	1.71	1.49

Furthermore, the semi-quantitative parameters, I_{003}/I_{104} and $c/(2\sqrt{6}a)$ ratios were calculated from the PXRD profiles and compared with the Ni fraction at the Li site obtained via the Rietveld refinement (Table 3). The ratio of integral intensities of the 003 and 104 reflections, after polarization, geometry and Lorentz corrections, and (ideally) corrections for absorption and temperature factors, can be considered as proportional to the squared ratio of the corresponding structure factors F_{003} and F_{104}:

$$\frac{I_{003}}{I_{104}} \sim \left(\frac{F_{003}}{F_{104}}\right)^2 = \left(\frac{f_{3a} - f_{3b} + 2f_O \cos(6\pi z_O)}{f_{3a} + f_{3b} + 2f_O \cos(8\pi z_O)}\right)^2 \quad (4)$$

Table 3. Semi-quantitative parameters for assessing the Li/Ni disorder in the NMC811 samples.

Sample	Ni in 3*a* Site, %	I_{003}/I_{104}	I_{003}/I_{104}, *po*-Corrected	I_{003}/I_{104}, *po*- and F_{000}-Corrected	$c/(2\sqrt{6}a)$
S1air	3.50(7)	1.290	1.185	1.202	1.00940
S1ox	2.45(6)	1.277	1.234	1.238	1.00875
S2air	5.02(7)	1.208	1.144	1.179	1.00876
S2ox	2.83(6)	1.312	1.227	1.191	1.00923
S3air	2.24(7)	1.382	1.301	1.243	1.00980
S3ox	2.61(6)	1.330	1.233	1.214	1.00953

Here, $f_{3a} = g_{Ni}f_{Ni} + (1 - g_{Ni})f_{Li}$, $f_{3b} = g_{Ni}f_{Ni} + (0.8 - g_{Ni})f_{Li} + 0.1(f_{Mn} + f_{Co})$, z_O—z-coordinate of the oxygen atom. This equation highlights the different nature of the 003 and 104 reflections: the former is the superlattice reflection arising from the layered ordering at the 3*a* and 3*b* sites with the structure factor proportional to the difference in scattering power at these positions, whereas the latter is the basic reflection of the parent rock-salt $Fm\bar{3}m$ structure with the structure factor correlating with F_{000} which counts the total number of electrons in the unit cell and depends on the overall Li and Ni content. The I_{003}/I_{104} intensity ratio also depends on the z-coordinate of the oxygen position, but it does not differ substantially if the violation of the layered ordering is relatively weak (Table 2) and can be safely neglected. The I_{003}/I_{104} intensity ratio has frequently been used for the assessment of the Li/Ni disorder [41–45].

Indeed, the I_{003}/I_{104} ratio generally correlates with the concentration of the cationic defects. The lowest (1.208) and highest (1.382) I_{003}/I_{104} ratios were obtained for the S3air and S2ox samples with the highest (5.02%) and lowest (2.24%) Ni fractions in the Li position, respectively. However, the dependence of I_{003}/I_{104} on the Ni content in the 3*a* site lacks systematic dependence for all six samples. Careful analysis of the PXRD patterns and residuals after the Rietveld refinement revealed that the origin of this discrepancy is in intensity distortion due to preferred orientation which is albeit small but different for all studied samples. The preferred intensity contribution was modeled with the March–Dollase function:

$$T_{hkl} = \left(\tau^2 cos^2\varphi_{hkl} + \frac{1}{\tau} sin^2\varphi_{hkl}\right)^{-3/2} \quad (5)$$

where T_{hkl} is the intensity correction factor, τ is the preferred orientation parameter obtained in the Rietveld refinement (Table 2) and φ_{hkl} is the angle between the reciprocal lattice vector $\boldsymbol{H}_{hkl}$ and preferred orientation direction (0° and 55° for the 003 and 104 reflections, respectively). After preferred orientation (*po*) correction, the I_{003}/I_{104} ratio follows the hyperbolic curve being plotted against the Ni content in the Li site, as expected from Equation (4) (Figure 2a). However, the air-annealed and oxygen-annealed samples cannot be distinguished from this dependence because it is governed by two factors simultaneously: $f_{3a} - f_{3b}$ difference and $f_{3a} + f_{3b}$ dependence on the overall Li and TM content. The two sets of samples with essentially different defect formation scenarios, as discussed above, can be discriminated by plotting the *po*-corrected I_{003}/I_{104} ratio normalized by squared F_{000} for each structure (Figure 2b), where the air- and oxygen-annealed samples follow two separate straight lines with different slopes.

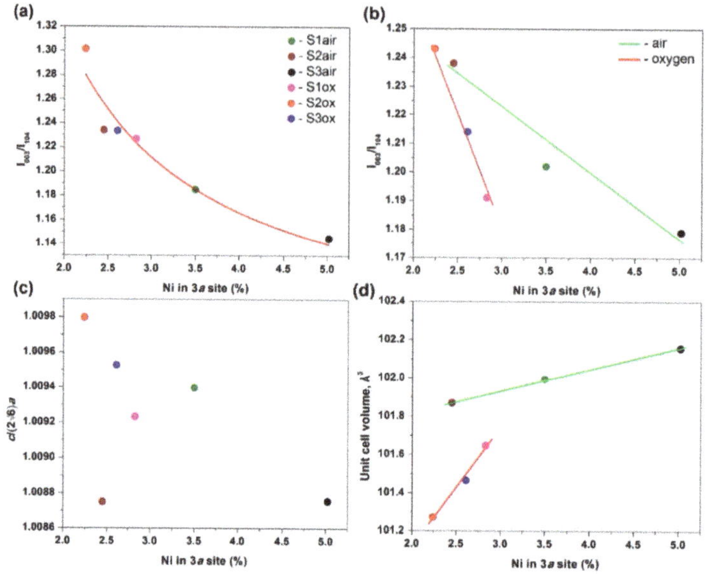

Figure 2. Correlation of semi-quantitative parameters with the Ni content in the Li site: (**a**) I_{003}/I_{104} ratio corrected for preferred orientation (red line show fitted hyperbolic curve, R^2 = 0.907); (**b**) I_{003}/I_{104} ratio corrected for preferred orientation and normalized by $(F_{000})^2$ (lines are guide to the eye, R^2 = 0.954 for the air-annealed samples, R^2 = 0.989 for the oxygen-annealed ones); (**c**) $c/(2\sqrt{6}a)$; (**d**) unit cell volume (lines are guide to the eye, R^2 = 0.999 for the air-annealed samples, R^2 = 0.969 for the oxygen-annealed ones).

To summarize, the I_{003}/I_{104} ratio can reflect the concentration of cationic point defects only qualitatively being sensitive even to small preferred orientation which is always present in the Ni-rich NMCs due to their layered structure. Sensible defect quantification with the I_{003}/I_{104} ratio requires knowing the defect formation model (as given with the Equations (1)–(3)), correction for preferred orientation and normalization per total scattering power of the unit cell. In addition, the I_{003}/I_{104} ratio cannot be used for a direct comparison of the materials if the PXRD patterns were measured with different experimental setups without corrections for polarization, Lorentz and geometry factors, and absorption corrections as well (that is particularly difficult as it is rarely measured and reported). The Rietveld method is the preferable technique for extracting the defect concentration from PXRD data.

The $c/(2\sqrt{6}a)$ parameter reflects the Li/Ni disorder only indirectly being a geometric measure of the deviation of the hexagonal unit cell of Ni-rich NMCs from their parent cubic rock-salt subcell in which $c/(2\sqrt{6}a) = 1$. Layered ordering of larger Li (r(Li^+) = 0.76 Å) and smaller Ni (r(Ni^{3+}) = 0.56 Å, r(Ni^{2+}) = 0.69 Å) at the *3a* and *3b* sites leads to anisotropic expansion of the cubic subcell along $\{111\}_c$ = [001]h direction and increase in $c/(2\sqrt{6}a)$ above 1. According to that, the largest $c/(2\sqrt{6}a)$ is observed for the less defect S2ox sample, whereas the most defect S3air sample demonstrates one of the smallest values. However, in general, the correlation between $c/(2\sqrt{6}a)$ and the Ni fraction in the Li position is surprisingly poor (Figure 2c). In fact, it is difficult to expect that the $c/(2\sqrt{6}a)$ ratio will behave monotonically with the Ni fraction at the *3a* site as this ratio is determined by the distribution of three atomic species with different ionic radii (Li^+, Ni^{2+} and Ni^{3+}) over two crystallographic sites. Unit cell volume appears to be more precise geometrical parameter which linearly correlates with the defect concentration according to Vegard law [46,47] (Figure 2d), but the slope of this linear dependence is essentially different for the air-annealed and oxygen-annealed samples ($V = p + qg_{Ni}$, V − unit cell volume ($Å^3$), g_{Ni} − Ni fraction in the 3*a* site (%); $p = 101.60(1)$, $q = 0.112(3)$, $R^2 = 0.999$ for the air-annealed samples; $p = 99.9(2)$, $q = 0.62(8)$, $R^2 = 0.969$ for the oxygen-annealed samples).

In order to perform a more local analysis of Ni^{2+}/Li^+ disorder, the crystal structure of all NMC811 samples was examined via the EDT technique. Additionally, single-crystal EDT experiments are not dependent on preferred orientation, in contrast to the PXRD data. For this, 3D-ED sets were acquired from three single crystals for each sample to provide relevant statistics. The crystal structure refinement from the 3D-ED data was performed in a similar scheme as described for PXRD data treatment. The refinement parameters and parameters of the Ni^{2+}/Li^+ mixing at the cationic sites are listed in Table S1. The structure refinement from the EDT data revealed reliability factors R_F within the range of 14–32% (Table S1). The higher R_F for the EDT data compared to PXRD ones are known to originate from the unavoidable dynamical scattering effects. As a result, electron diffraction conditions in the 3D-ED deviate from kinematic approximation, commonly applied while treating the diffraction data, even for 100–200 nm crystallites from which the 3D-ED data were collected. Nevertheless, the obtained reliability factors are generally considered credible for this method [36,37].

The obtained results from EDT data demonstrate the absence of Li^+ at the 3*b* sites for air-annealed samples, thus reaffirming the results of Rietveld refinement. For all three air-annealed samples, g_{Li} at 3*b* sites turned out to have zero value within one standard deviation. Regarding the oxygen-annealed materials, for S3ox sample g_{Li} at the 3*b* sites estimated with 3D-ED is very similar to that refined from PXRD data within the statistical error, while for S1ox and S2ox samples 3D-ED pointed to slightly smaller Li^+ concentration at the 3*b* site than those obtained from Rietveld refinement. However, for all oxygen-annealed materials g_{Li} at 3*b* sites is statistically significant, which means that switching the annealing atmosphere from air to oxygen changes the defect formation mechanism. The comparison of the Ni fraction in the Li site determined with the Rietveld refinement from PXRD data and from EDT demonstrates a solid correlation (Figure 3) that renders the EDT technique a reliable method for the determination of cationic point defects in the crystal structure of layered oxides. The most important issue is the validity of kinematical structure refinement from EDT data due to the strong dynamic scattering effects, which can be partially mitigated by selecting sufficiently thin crystals for data collection but could also be more rigorously addressed in the dynamical effects are reduced of fully taken into account [48–51]. Although being not superior compared to the Rietveld refinement (at least within kinematical approximation) and intrinsically lower in representativity, EDT might become indispensable if the Ni/Li disorder should be determined in the electrochemically-treated electrodes which often demonstrate poor PXRD patterns deteriorated by amorphous additives, current collector and multiphase nature.

Although both PXRD and EDT provide a reliable characterization of the cation disorder, they are not capable of delivering local information, such as defect concentration at the surface and in the bulk of the NMC crystallites. Atomic-resolution HAADF-STEM imaging has been frequently used for assessing local Ni^{2+}/Li^+ disorder, but only in a qualitative manner. The [010] and [$\bar{1}$10] HAADF-STEM images, in which the Li and TM slabs can be easily distinguished, were interpreted visually, sometimes supporting the conclusions by plotting the intensity profiles. The local structure of the disordered regions was analyzed by estimation of the brightness of the dots, associated with Ni ions migrated to Li slabs, and followed by tentative assignment of the local areas to layered, spinel, or rock-salt structure types [15,27–31]. The development of state-of-the-art techniques, aimed at the quantitative analysis of high-resolution STEM images, based on the application of statistical parameter estimation theory [33,52–54], makes it possible to directly quantify the number of atoms in each projected atomic column in the HAADF-STEM image. Until recently, the HAADF-STEM image processing was applied predominantly to the metallic nanoparticles [55], while within the present work we attempted to apply such an approach to cation disordering in Ni-rich NMCs.

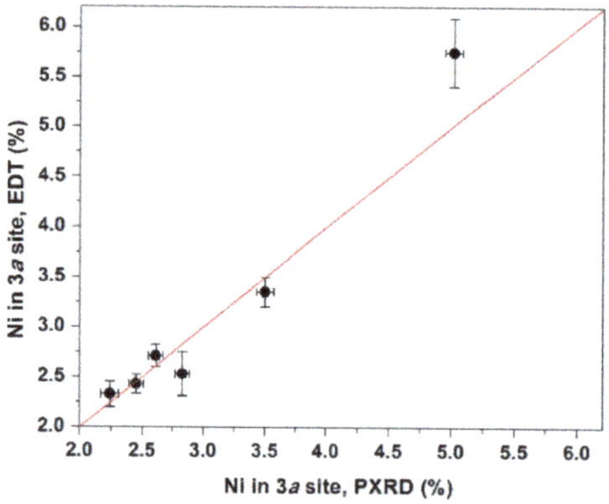

Figure 3. Correlation of the Ni fraction at the Li site as refined from PXRD and EDT data. The bars indicate standard deviations obtained in each method.

Several assumptions were made to apply the statistical parameter estimation theory. First, the scattering power of Ni, Mn and Co was considered equal as their atomic numbers are close to each other, meaning that the TMs were considered as one type of atom. Second, due to the strong dependence of HAADF signal on average atomic number Z along the atomic column ($\approx Z^n$, where $1.4 \leq n \leq 2$) [56], the Li-ions impact on the scattering cross-sections can be considered as negligible. Thus, all the atomic columns in the HAADF-STEM image of Ni-rich NMCs virtually consist of TM ions exclusively and only TM cations were counted while performing calculations. Oxygen atoms weakly contribute to the HAADF-STEM images, but they do not mix with the cationic columns if the structure is viewed along [010]/[100] direction of the $R\bar{3}m$ structure selected for imaging. This direction also provides well-separated images of the atomic columns corresponding to the 3*a* and 3*b* sites.

For further investigation, the S1air, S1ox and S3ox NMC811 samples were sorted to demonstrate the correlation between the distribution of disordered cation and synthesis conditions. Such choice is attributed to the similar synthesis route for these samples via hydroxide co-precipitation from TM sulfates with the difference in the pH at the first step of the synthesis procedure and annealing atmosphere at the second stage. In addition, the S3air sample, unlike the others prepared by carbonate routine, was selected as a comparative example. The [010] HAADF-STEM images were collected for S1air, S1ox, S3air, and S3ox NMC811 samples (Figure 4), followed by the quantitative analysis of a fraction of the Ni ions, migrated to the Li^+ 3*a* sites via statistical parameter estimation theory, employed into the StatSTEM software [33]. Figure 4b,e,h,k demonstrates the color-coded number of TM atoms in the atomic columns of the 3*a* and 3*b* sites. The parameter $P = \frac{N_{3a}}{N_{3a}+N_{3b}}$ (*N* is the number of TM cations in the atomic column of the corresponding site) was retrieved from the atom number map using a dedicated clusterization procedure described in more detail in Supplementary Materials. This parameter is related to the ratio of the TM occupancies at the 3*a* and 3*b* sites as $\frac{g(TM)_{3a}}{g(TM)_{3b}} = \frac{P}{1-P}$. The clusterization color maps, shown in Figure 4c,f,i,l illustrate the spatial distribution of regions with various disordering levels. The amount of disorder in each sample is generally decreasing from the edge to the bulk with the major part located at the surface. It is worth noting that Ni in Li sites is distributed non-uniformly, forming domains rather than a continuous layer with definite thickness, as was stated in earlier works. This observation is quite unexpected and confronts the conjecture on a formation of a "blocking" surface layer that impedes the Li diffusion due to the high Ni fraction in the Li site. In fact, the discontinuous nature of the disordered surface region might create percolated diffusion pathways, thus maintaining high Li permeability. Therefore, the precise structure of the "blocking" layer deserves further in-depth investigations.

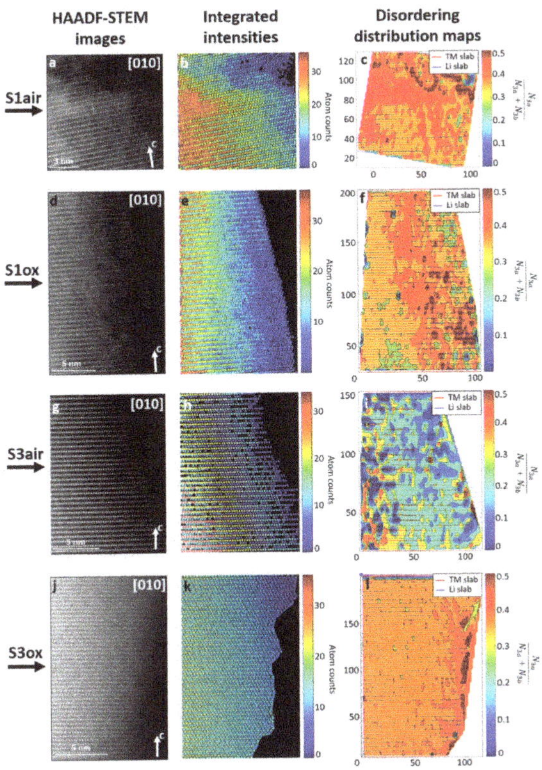

Figure 4. Quantitative analysis of atomic-resolution HAADF-STEM images of S1air, S1ox, S3air, and S3ox NMC811 samples with statistical parameter estimation theory. For each sample [010] HAADF-STEM images (left column, **a**,**d**,**g**,**j**), corresponding color-coded number of TM atoms in each atomic column (middle column, **b**,**e**,**h**,**k**), and the Ni ions cation disordering distribution color map of the regions with various disordering levels in the near-surface region (right column, **c**,**f**,**i**,**l**) are shown.

Concerning the particular disordering level at different regions, the maps of hydroxide-prepared samples S1air, S1ox and S3ox clearly show that, despite the relative closeness of the disordering level, their distribution varies significantly from the S1air sample, where defects are propagating towards the bulk of the crystal, to the S3ox, where defects are segregated at the 20–25 Å surface layer. Meanwhile, similar maps for the carbonate-prepared S3air sample demonstrate the uniform distribution of distinct disordered regions in both surface and bulk, as opposed to the hydroxide-prepared ones with a gradual changing of disordering level. Thus, the quantitative analysis demonstrates that even the slightest change in synthesis conditions leads to the significant redistribution of the cation disordering.

Indeed, the proposed method opens doors for the proper monitoring of the disordered cation concentration and their localization at the atomic scale, which could help to shed light on the impact of the cation interslab exchange on electrochemical properties of cathode materials. However, like every method, quantitative HAADF-STEM imaging also possesses certain limitations and should be applied carefully in order not to avoid artifacts and overinterpretations. Foremost, the HAADF-STEM technique requires a relatively high electron dose to maintain a suitable signal-to-noise ratio, but prolonged beam exposure can lead to the formation of beam-induced cationic point defects formation in layered NMCs cathode materials. These defects are similar to those formed during synthesis and/or battery operation that may result in misleading image interpretation [57,58]. The last but not the least problem of the quantitative HAADF-STEM imaging is representativity, which is here even more acute than for the EDT method since no more than a few dozens of nm^2 space can be scanned in a single image. The depth of field of the electron probe, limited to $\approx$100–150 nm, and strong $\bar{e}$-beam interaction with the matter also dictates the possibility of studying only the near-surface area of the crystallite, while its deeper inner part remains virtually inaccessible. This disadvantage, however, can be mitigated by the preparation of uniformly thin slices with a focused ion beam (FIB) technique, but mobility and migration probability of the Ni and Li cations under high energy ion beam must be perfectly understood first.

4. Conclusions

The comprehensive quantitative and semi-quantitative analysis of Ni^{2+}/Li^{+} cation disordering in Ni-rich NMC ($LiNi_{0.8}Mn_{0.1}Co_{0.1}O_2$) samples were performed by PXRD, 3D-ED (EDT), and quantitative HAADF-STEM on various spatial scales. Rietveld refinement from PXRD data provides primary information on the cation disorder schemes, which strongly depend on the synthesis conditions. Estimation of defect concentrations in a semi-quantitative manner is also possible using the I_{003}/I_{104} integral intensity ratio but only after correction for preferred orientation, whereas the $c/(2\sqrt{6}a)$ ratio of pseudocubic subcell parameters does not quantitatively correlate with the cation disorder and must be used with extreme care. EDT appears to measure the cation disorder reliably correlating well with the Rietveld refinement results while delivering the information about defects distribution on the level of separate crystallites. The work highlights a new computational approach for the cation disordering study, based on quantitative analysis of high-resolution HAADF-STEM images, which provides an unprecedented opportunity to identify and distinguish regions with various Ni^{2+}/Li^{+} disorder levels. The proposed technique could potentially serve as an efficient tool for the investigation of Ni^{2+}/Li^{+} defect structure at the atomic scale, providing detailed information about its formation, localization and evolution during electrochemical cycling.

Supplementary Materials: The following are available online at https://www.mdpi.com/2073-8994/13/9/1628/s1, Figure S1–S6: Experimental, calculated and difference PXRD profiles after the Rietveld refinement of the NMC811 structure in the S1air, S1ox, S2air, S2ox, S3air, S3ox samples. Table S1: Results of the 3D ED study of Ni^{2+}/Li^{+} anti-site defects. Table S1. Results of the 3D-ED study of Ni^{2+}/Li^{+} anti-site defects. Clusterization procedure description.

Author Contributions: Conceptualization, A.M.A.; methodology, A.A.S., E.D.O., S.A.A. and A.M.A.; validation, E.D.O. and A.A.S.; formal analysis, A.A.S. and E.D.O.; investigation, A.M.A., A.A.S., E.D.O. and A.V.M.; writing—original draft preparation, E.D.O.; writing—review and editing, A.M.A. and A.A.S.; supervision, A.M.A.; funding acquisition, A.M.A. All authors have read and agreed to the published version of the manuscript.

Funding: This research was funded by the Russian Science Foundation, grant 20-13-00233.

Institutional Review Board Statement: Not applicable.

Informed Consent Statement: Not applicable.

Data Availability Statement: Not applicable.

Acknowledgments: Access to the Transmission Electron Microscopy (TEM) facilities was granted by the Advanced Imaging Core Facility (AICF) of Skoltech.

Conflicts of Interest: The authors declare no conflict of interest.

References

1. Whittingham, M.S. Ultimate limits to intercalation reactions for lithium batteries. *Chem. Rev.* **2014**, *114*, 11414–11443. [CrossRef]
2. Goodenough, J.B.; Kim, Y. Challenges for rechargeable Li batteries. *Chem. Mater.* **2009**, *22*, 587–603. [CrossRef]
3. Manthiram, A.; Knight, J.C.; Myung, S.-T.; Oh, S.-M.; Sun, Y.-K. Nickel-rich and lithium-rich layered oxide cathodes: Progress and perspectives. *Adv. Energy Mater.* **2015**, *6*. [CrossRef]
4. Delmas, C.; Fouassier, C.; Hagenmuller, P. Structural classification and properties of the layered oxides. *Physica* **1980**, *99*, 81–85. [CrossRef]
5. Shannon, R.D. Revised effective ionic radii and systematic studies of interatomic distances in halides and chalcogenides. *Acta Crystallogr. Sect. A Cryst. Phys. Diffr. Theor. Gen. Crystallogr.* **1976**, *32*, 751–767. [CrossRef]
6. Gao, A.; Sun, Y.; Zhang, Q.; Zheng, J.; Lu, X. Evolution of Ni/Li antisites under the phase transition of a layered LiNi1/3Co1/3Mn1/3O2 cathode. *J. Mater. Chem. A* **2020**, *8*, 6337–6348. [CrossRef]
7. Zheng, J.; Ye, Y.; Liu, T.; Xiao, Y.; Wang, C.; Wang, F.; Pan, F. Ni/Li disordering in layered transition metal oxide: Electrochemical impact, origin, and control. *Acc. Chem. Res.* **2019**, *52*, 2201–2209. [CrossRef]
8. Xu, J.; Lin, F.; Doeff, M.; Tong, W. A review of Ni-based layered oxides for rechargeable Li-ion batteries. *J. Mater. Chem. A* **2016**, *5*, 874–901. [CrossRef]
9. Kang, K.; Meng, Y.S.; Bréger, J.; Grey, C.P.; Ceder, G. Electrodes with high power and high capacity for rechargeable lithium batteries. *Science* **2006**, *311*, 977–980. [CrossRef]
10. Yu, H.; Qian, Y.; Otani, M.; Tang, D.; Guo, S.; Zhu, Y.; Zhou, H. Study of the lithium/nickel ions exchange in the layered LiNi0.42Mn0.42Co0.16O2 cathode material for lithium ion batteries: Experimental and first-principles calculations. *Energy Environ. Sci.* **2014**, *7*, 1068–1078. [CrossRef]
11. Zhang, S.S. Problems and their origins of Ni-rich layered oxide cathode materials. *Energy Storage Mater.* **2019**, *24*, 247–254. [CrossRef]
12. Bianchini, M.; Roca-Ayats, M.; Hartmann, P.; Brezesinski, T.; Janek, J. There and back again—The journey of LiNiO 2 as a cathode active material. *Angew. Chem. Int. Ed.* **2018**, *58*, 10434–10458. [CrossRef]
13. Yan, P.; Nie, A.; Zheng, J.; Zhou, Y.; Lu, D.; Zhang, X.; Xu, R.; Belharouak, I.; Zu, X.; Xiao, J.; et al. Evolution of lattice structure and chemical composition of the surface reconstruction layer in Li1.2Ni0.2Mn0.6O2 cathode material for lithium ion batteries. *Nano Lett.* **2014**, *15*, 514–522. [CrossRef]
14. Yan, P.; Zheng, J.; Lv, D.; Wei, Y.; Zheng, J.; Wang, Z.; Kuppan, S.; Yu, J.; Luo, L.; Edwards, D.; et al. Atomic-resolution visualization of distinctive chemical mixing behavior of Ni, Co, and Mn with Li in layered lithium transition-metal oxide cathode materials. *Chem. Mater.* **2015**, *27*, 5393–5401. [CrossRef]
15. Lin, F.; Markus, I.M.; Nordlund, D.; Weng, T.-C.; Asta, M.D.; Xin, H.; Doeff, M. Surface reconstruction and chemical evolution of stoichiometric layered cathode materials for lithium-ion batteries. *Nat. Commun.* **2014**, *5*, 3529. [CrossRef]
16. Li, T.; Yuan, X.-Z.; Zhang, L.; Song, D.; Shi, K.; Bock, C. Degradation mechanisms and mitigation strategies of nickel-rich NMC-based lithium-ion batteries. *Electrochem. Energy Rev.* **2019**, *3*, 43–80. [CrossRef]
17. Liu, W.; Oh, P.; Liu, X.; Lee, M.-J.; Cho, W.; Chae, S.; Kim, Y.; Cho, J. Nickel-rich layered lithium transition-metal oxide for high-energy lithium-ion batteries. *Angew. Chem. Int. Ed.* **2015**, *54*, 4440–4457. [CrossRef]
18. Zheng, J.X.; Liu, T.C.; Hu, Z.X.; Wei, Y.; Song, X.H.; Ren, Y.; Wang, W.D.; Rao, M.M.; Lin, Y.; Chen, Z.H.; et al. Tuning of thermal stability in layered Li(NixMnyCoz)O2. *J. Am. Chem. Soc.* **2016**, *138*, 13326–13334. [CrossRef] [PubMed]
19. Meng, Y.; Ceder, G.; Grey, C.; Yoon, W.-S.; Jiang, M.; Breger, J.; Shao-Horn, Y. Cation ordering in layered O3 Li[NixLi1/3−2 x/3Mn2/3-x/3]O2 ($0 \leq x \leq 1/2$) compounds. *Chem. Mater.* **2005**, *17*, 2386–2394. [CrossRef]
20. Sun, G.; Yin, X.; Yang, W.; Song, A.; Jia, C.; Yang, W.; Du, Q.; Ma, Z.; Shao, G. The effect of cation mixing controlled by thermal treatment duration on the electrochemical stability of lithium transition-metal oxides. *Phys. Chem. Chem. Phys.* **2017**, *19*, 29886–29894. [CrossRef] [PubMed]
21. Kong, F.; Liang, C.; Wang, L.; Zheng, Y.; Perananthan, S.; Longo, R.C.; Ferraris, J.P.; Kim, M.; Cho, K. Kinetic stability of bulk LiNiO2 and surface degradation by oxygen evolution in LiNiO2 based cathode materials. *Adv. Energy Mater.* **2018**, *9*. [CrossRef]
22. Tian, C.; Nordlund, D.; Xin, H.; Xu, Y.; Liu, Y.; Sokaras, D.; Lin, F.; Doeff, M.M. Depth-dependent redox behavior of LiNi0.6Mn0.2Co0.2O2. *J. Electrochem. Soc.* **2018**, *165*, A696–A704. [CrossRef]
23. Zou, L.; Zhao, W.; Liu, Z.; Jia, H.; Zheng, J.; Wang, G.; Yang, Y.; Zhang, J.-G.; Wang, C. Revealing cycling rate-dependent structure evolution in ni-rich layered cathode materials. *ACS Energy Lett.* **2018**, *3*, 2433–2440. [CrossRef]
24. Xiao, Y.; Liu, T.; Liu, J.; He, L.; Chen, J.; Zhang, J.; Luo, P.; Lu, H.; Wang, R.; Zhu, W.; et al. Insight into the origin of lithium/nickel ions exchange in layered Li(NixMnyCoz)O2 cathode materials. *Nano Energy* **2018**, *49*, 77–85. [CrossRef]
25. Yin, L.; Mattei, G.S.; Li, Z.; Zheng, J.; Zhao, W.; Omenya, F.; Fang, C.; Li, W.; Li, J.; Xie, Q.; et al. Extending the limits of powder diffraction analysis: Diffraction parameter space, occupancy defects, and atomic form factors. *Rev. Sci. Instrum.* **2018**, *89*, 093002. [CrossRef] [PubMed]
26. Yin, L.; Li, Z.; Mattei, G.S.; Zheng, J.; Zhao, W.; Omenya, F.; Fang, C.; Li, W.; Li, J.; Xie, Q.; et al. Thermodynamics of antisite defects in layered NMC cathodes: Systematic insights from high-precision powder diffraction analyses. *Chem. Mater.* **2019**, *32*, 1002–1010. [CrossRef]
27. Hwang, S.; Su, D. TEM studies on electrode materials for secondary ion batteries. *Inorg. Battery Mater.* **2019**, *311*, 1–27. [CrossRef]

28. Zhu, J.; Sharifi-Asl, S.; Garcia, J.C.; Iddir, H.H.; Croy, J.R.; Shahbazian-Yassar, R.; Chen, G. Atomic-level understanding of surface reconstruction based on Li[NixMnyCo1−x−y]O2 single-crystal studies. *ACS Appl. Energy Mater.* **2020**, *3*, 4799–4811. [CrossRef]
29. Kim, U.-H.; Park, G.-T.; Conlin, P.; Ashburn, N.; Cho, K.; Yu, Y.-S.; Shapiro, D.A.; Maglia, F.; Kim, S.-J.; Lamp, P.; et al. Cation ordered Ni-rich layered cathode for ultra-long battery life. *Energy Environ. Sci.* **2021**, *14*, 1573–1583. [CrossRef]
30. Jung, S.-K.; Gwon, H.; Hong, J.; Park, K.-Y.; Seo, D.-H.; Kim, H.; Hyun, J.; Yang, W.; Kang, K. Understanding the degradation mechanisms of LiNi0.5 Co0.2 Mn0.3 O2 cathode material in lithium ion batteries. *Adv. Energy Mater.* **2013**, *4*. [CrossRef]
31. Schweidler, S.; De Biasi, L.; Garcia, G.; Mazilkin, A.; Hartmann, P.; Brezesinski, T.; Janek, J. Investigation into mechanical degradation and fatigue of high-Ni NCM cathode material: A long-term cycling study of full cells. *ACS Appl. Energy Mater.* **2019**, *2*, 7375–7384. [CrossRef]
32. Van Aert, S.; De Backer, A.; Martinez, G.T.; Dekker, A.D.; Van Dyck, D.; Bals, S.; Van Tendeloo, G. Advanced electron crystallography through model-based imaging. *IUCrJ* **2016**, *3*, 71–83. [CrossRef]
33. De Backer, A.; Bos, K.H.W.V.D.; Broek, W.V.D.; Sijbers, J.; Van Aert, S. StatSTEM: An efficient approach for accurate and precise model-based quantification of atomic resolution electron microscopy images. *Ultramicroscopy* **2016**, *171*, 104–116. [CrossRef]
34. Gemmi, M.; Mugnaioli, E.; Gorelik, T.E.; Kolb, U.; Palatinus, L.; Boullay, P.; Hovmöller, S.; Abrahams, J.P. 3D electron diffraction: The nanocrystallography revolution. *ACS Cent. Sci.* **2019**, *5*, 1315–1329. [CrossRef]
35. Vainshtein, B.K.; Zvyagin, B.B.; Avilov, A.S. *Diffraction Structure Analysis in Electron—Diffraction Techniques*; Cowley, J.M., Ed.; Oxford University Press: Oxford, UK, 1992; pp. 216–312.
36. Hadermann, J.; Abakumov, A.M. Structure solution and refinement of metal-ion battery cathode materials using electron diffraction tomography. *Acta Crystallogr. Sect. B Struct. Sci. Cryst. Eng. Mater.* **2019**, *75*, 485–494. [CrossRef] [PubMed]
37. Karakulina, O.M.; Khasanova, N.R.; Drozhzhin, O.A.; Tsirlin, A.A.; Hadermann, J.; Antipov, E.V.; Abakumov, A.M. Antisite disorder and bond valence compensation in Li2FePO4F cathode for Li-Ion batteries. *Chem. Mater.* **2016**, *28*, 7578–7581. [CrossRef]
38. Mugnaioli, E. Closing the gap between electron and X-ray crystallography. *Acta Crystallogr. Sect. B Struct. Sci. Cryst. Eng. Mater.* **2015**, *71*, 737–739. [CrossRef] [PubMed]
39. Petříček, V.; Dušek, M.; Palatinus, L. Crystallographic computing system JANA2006: General features. *Z. Krist. Cryst. Mater.* **2014**, *229*, 345–352. [CrossRef]
40. Palatinus, L.; Brázda, P.; Jelínek, M.; Hrda, J.; Steciuk, G.; Klementova, M. Specifics of the data processing of precession electron diffraction tomography data and their implementation in the program PETS2.0. *Acta Crystallogr. Sect. B Struct. Sci. Cryst. Eng. Mater.* **2019**, *75*, 512–522. [CrossRef]
41. Nam, K.-W.; Bak, S.-M.; Hu, E.; Yu, X.; Zhou, Y.; Wang, X.; Wu, L.; Zhu, Y.; Chung, K.Y.; Yang, X.-Q. Combining in situ synchrotron X-ray diffraction and absorption techniques with transmission electron microscopy to study the origin of thermal instability in overcharged cathode materials for Lithium-Ion batteries. *Adv. Funct. Mater.* **2012**, *23*, 1047–1063. [CrossRef]
42. Morales, J.; Vicente, C.P.; Tirado, J.L. Cation distribution and chemical deintercalation of Li1-xNi1+xO2. *Mater. Res. Bull.* **1990**, *25*, 623–630. [CrossRef]
43. Ohzuku, T.; Ueda, A.; Nagayama, M. Electrochemistry and structural chemistry of LiNiO2 (R3m) for 4 volt secondary lithium Cells. *J. Electrochem. Soc.* **1993**, *140*, 1862–1870. [CrossRef]
44. Hua, W.; Zhang, J.; Zheng, Z.; Liu, W.; Peng, X.; Guo, X.-D.; Zhong, B.; Wang, Y.-J.; Wang, X. Na-doped Ni-rich LiNi0.5Co0.2Mn0.3O2 cathode material with both high rate capability and high tap density for lithium ion batteries. *Dalton Trans.* **2014**, *43*, 14824–14832. [CrossRef]
45. Zhang, X.; Jiang, W.; Mauger, A.; Qilu, A.; Gendron, F.; Julien, C. Minimization of the cation mixing in Li1+x(NMC)1−xO2 as cathode material. *J. Power Sources* **2010**, *195*, 1292–1301. [CrossRef]
46. Vegard, L. Die konstitution der mischkristalle und die raumfüllung der atome. *Z. Phys.* **1921**, *5*, 17–26. [CrossRef]
47. Denton, A.R.; Ashcroft, N.W. Vegard's law. *Phys. Rev. A* **1991**, *43*, 3161–3164. [CrossRef] [PubMed]
48. Mugnaioli, E.; Gemmi, M. Single-crystal analysis of nanodomains by electron diffraction tomography: Mineralogy at the order-disorder borderline. *Z. Kristallogr. Cryst. Mater.* **2018**, *233*, 163–178. [CrossRef]
49. Palatinus, L.; Jacob, D.; Cuvillier, P.; Klementova, M.; Sinkler, W.; Marks, L.D. Structure refinement from precession electron diffraction data. *Acta Crystallogr. Sect. A Found. Crystallogr.* **2013**, *69*, 171–188. [CrossRef]
50. Palatinus, L.; Petříček, V.; Corrêa, C.A. Structure refinement using precession electron diffraction tomography and dynamical diffraction: Theory and implementation. *Acta Crystallogr. Sect. A Found. Adv.* **2015**, *71*, 235–244. [CrossRef]
51. Palatinus, L.; Corrêa, C.A.; Steciuk, G.; Jacob, D.; Roussel, P.; Boullay, P.; Klementová, M.; Gemmi, M.; Kopeček, J.; Domeneghetti, M.C.; et al. Structure refinement using precession electron diffraction tomography and dynamical diffraction: Tests on experimental data. *Acta Cryst. B* **2015**, *71*, 740–751. [CrossRef]
52. LeBeau, J.M.; Findlay, S.D.; Allen, L.J.; Stemmer, S. Standardless atom counting in scanning transmission electron microscopy. *Nano Lett.* **2010**, *10*, 4405–4408. [CrossRef] [PubMed]
53. Van Aert, S.; Batenburg, K.J.; Rossell, M.D.; Erni, R.; Van Tendeloo, G. Three-dimensional atomic imaging of crystalline nanoparticles. *Nature* **2011**, *470*, 374–377. [CrossRef] [PubMed]
54. Bals, S.; Van Aert, S.; Van Tendeloo, G.; Brande, D.A. Statistical estimation of atomic positions from exit wave reconstruction with a precision in the picometer range. *Phys. Rev. Lett.* **2006**, *96*, 096106. [CrossRef]

55. De Backer, A.; Jones, L.; Lobato, I.; Altantzis, T.; Goris, B.; Nellist, P.D.; Bals, S.; Van Aert, S. Three-dimensional atomic models from a single projection using Z-contrast imaging: Verification by electron tomography and opportunities. *Nanoscale* **2017**, *9*, 8791–8798. [CrossRef] [PubMed]
56. Guzzinati, G.; Altantzis, T.; Batuk, M.; De Backer, A.; Lumbeeck, G.; Samaee, V.; Batuk, D.; Idrissi, H.; Hadermann, J.; Van Aert, S.; et al. Recent advances in transmission electron microscopy for materials science at the EMAT lab of the University of Antwerp. *Materials* **2018**, *11*, 1304. [CrossRef]
57. Lin, F.; Markus, I.M.; Doeff, M.; Xin, H.L. Chemical and structural stability of lithium-ion battery electrode materials under electron beam. *Sci. Rep.* **2014**, *4*, 5694. [CrossRef]
58. Hobbs, L.W. Electron-beam sensitivity in inorganic specimens. *Ultramicroscopy* **1987**, *23*, 339–344. [CrossRef]

symmetry MDPI

Correction

Correction: Rauch et al. New Features in Crystal Orientation and Phase Mapping for Transmission Electron Microscopy. *Symmetry* 2021, *13*, 1675

Edgar F. Rauch [1,*], Patrick Harrison [1], Xuyang Zhou [2], Michael Herbig [2], Wolfgang Ludwig [3] and Muriel Véron [1]

1 SIMAP, Grenoble INP, Université Grenoble Alpes, CNRS, 38000 Grenoble, France; patrick.harrison@simap.grenoble-inp.fr (P.H.); muriel.veron@grenoble-inp.fr (M.V.)
2 Max-Planck-Institut für Eisenforschung GmbH, Max-Planck-Str. 1, 40237 Düsseldorf, Germany; x.zhou@mpie.de (X.Z.); m.herbig@mpie.de (M.H.)
3 MATEIS, INSA Lyon, Université Lyon I, CNRS UMR 5510, 69621 Villeurbanne, France; ludwig@esrf.fr
* Correspondence: edgar.rauch@grenoble-inp.fr

1. Addition of Authors

The authors wish to make the following corrections to this paper [1].

X.Z., M.H. and W.L. were not included as authors in the original publication. The corrected Author Contributions Statement appears here. The authors apologize for any inconvenience caused and state that the scientific conclusions are unaffected. The original publication has also been updated.

2. Author Contributions Statement

Xuyang Zhou: Specimen preparation, Data acquisition
Michael Herbig: Conceptualization, Methodology, Funding acquisition
Wolfgang Ludwig: Conceptualization, Methodology, Funding acquisition

Author Contributions: Conceptualization, methodology, E.F.R., W.L. and M.H.; software, E.F.R. and P.H.; validation, E.F.R., P.H. and M.V.; formal analysis, E.F.R. and P.H.; investigation, E.F.R., X.Z. and P.H.; resources, E.F.R., M.H. and M.V.; writing—original draft preparation, E.F.R. and P.H.; writing—review and editing, E.F.R. and P.H.; visualization, E.F.R., X.Z. and P.H.; supervision, E.F.R., M.H. and M.V.; project administration, E.F.R.; funding acquisition, E.F.R., W.L., M.H. and M.V. All authors have read and agreed to the published version of the manuscript.

Funding: This research was partly funded by Agence National de la Recherche (ANR), grant number ANR-19-CE42-0017. M.H. acknowledges funding by the German Research Foundation (DFG) for funding via project HE 7225/11-1. The work has benefited from characterization equipment of the Grenoble INP-CMTC platform supported by the Centre of Excellence of Multifunctional Architectured Materials 'CEMAM' n°ANR-10-LABX-44-01 funded by the Investments for the Future programme.

Reference

1. Rauch, E.F.; Harrison, P.; Zhou, X.; Herbig, M.; Ludwig, W.; Véron, M. New Features in Crystal Orientation and Phase Mapping for Transmission Electron Microscopy. *Symmetry* **2021**, *13*, 1675. [CrossRef]

symmetry

MDPI

Article

A Comparison of Structure Determination of Small Organic Molecules by 3D Electron Diffraction at Cryogenic and Room Temperature

Taimin Yang [1], Steve Waitschat [2], Andrew Kentaro Inge [1], Norbert Stock [2], Xiaodong Zou [1,*] and Hongyi Xu [1,*]

[1] Department of Materials and Environmental Chemistry, Stockholm University, 106 91 Stockholm, Sweden; taimin.yang@mmk.su.se (T.Y.); andrew.inge@mmk.su.se (A.K.I.)

[2] Institut für Anorganische Chemie, Christian-Albrechts-Universität zu Kiel, 241 18 Kiel, Germany; swaitschat@ac.uni-kiel.de (S.W.); stock@ac.uni-kiel.de (N.S.)

* Correspondence: xzou@mmk.su.se (X.Z.); hongyi.xu@mmk.su.se (H.X.)

Abstract: 3D electron diffraction (3D ED), also known as micro-crystal electron diffraction (MicroED), is a rapid, accurate, and robust method for structure determination of submicron-sized crystals. 3D ED has mainly been applied in material science until 2013, when MicroED was developed for studying macromolecular crystals. MicroED was considered as a cryo-electron microscopy method, as MicroED data collection is usually carried out in cryogenic conditions. As a result, some researchers may consider that 3D ED/MicroED data collection on crystals of small organic molecules can only be performed in cryogenic conditions. In this work, we determined the structure for sucrose and azobenzene tetracarboxylic acid (H_4ABTC). The structure of H_4ABTC is the first crystal structure ever reported for this molecule. We compared data quality and structure accuracy among datasets collected under cryogenic conditions and room temperature. With the improvement in data quality by data merging, it is possible to reveal hydrogen atom positions in small organic molecule structures under both temperature conditions. The experimental results showed that, if the sample is stable in the vacuum environment of a transmission electron microscope (TEM), the data quality of datasets collected under room temperature is at least as good as data collected under cryogenic conditions according to various indicators (resolution, $I/\sigma(I)$, $CC_{1/2}$ (%), R_1, R_{int}, ADRA).

Keywords: 3D ED; MicroED; electron crystallography; structure determination; cryogenic TEM; room temperature

1. Introduction

Electrons are ideal radiation sources for diffraction and imaging experiments on submicron-sized crystals because of their strong interaction with matter. However, the strong interaction brings severe radiation damage at the same time. It is difficult to obtain high-resolution transmission electron microscope (HRTEM) images from organic crystals because acquiring high signal-to-noise ratio (SNR) HRTEM images requires a higher electron dose [1]. Electron diffraction, on the other hand, requires a much lower electron dose to achieve atomic resolution and it is suitable for studying organic crystals. In 1976, Dorset and Hauptmann applied direct methods for ab initio phasing of electron diffraction data from organic crystals [2]. In the early 1990s, Dorset et al. demonstrated ab initio structure determination of a wide range of organic crystals using electron diffraction data [3,4]. In all cases, crystal structure models were obtained from one or only a few 2D zone-axis electron diffraction patterns. With the development of digitized detectors, the collection of 3D electron diffraction data from a single submicron-sized crystal for structure determination became possible. Kolb et al. and Zou et al. independently developed 3D electron diffraction methods, named automated electron diffraction tomography (ADT) [5,6] and rotation electron diffraction (RED) [7,8], respectively. In 2010, Kolb et al. demonstrated that it is possible to analyze structures of organics using ADT [9]. Abrahams et al. determined structures of two organic pharmaceutical compounds by continuous rotation 3D electron diffraction using a hybrid pixel detector in 2016 [10]. This was the first time that such a hybrid pixel detector was used for collecting 3D ED data, which has the advantages of high sensitivity, low noise and, short readout time. Later, several articles [11–17] described the routine of rapid structure determination for small organic molecules using 3D ED. Palatinus et al. showed that it was possible to determine hydrogen positions in both organic (paracetamol) and inorganic crystals by employing dynamical refinements [18]. In 2019, they determined the molecular absolute configuration for a pharmaceutical compound [19]. Electron diffraction methods were also applied for structure determination of micron- and nano-sized protein crystals under cryogenic conditions by several groups [20–25]. 3D ED can also be performed in nanobeam mode [26,27] and combined with 4D STEM [27] to study small organic molecules and protein crystals. To alleviate radiation damage, researchers showed that the structure model can be improved by merging small-wedge 3D ED datasets collected from multiple crystals [28]. Furthermore, recent results showed that it is even possible to solve unknown protein structures (phased from a homologue less than 40% in sequence identity or having different conformation compared to the molecular replacement model) by MicroED [29,30], and visualize ligand binding interactions with electron diffraction datasets [31].

With the achievements mentioned above, 3D ED is becoming a standalone and reliable method for structure determination. However, some established protocols and methods were focused on data collection of macromolecules and organic molecules in cryogenic conditions, and were integrated into the existing cryo-EM sample preparation workflow. MicroED was introduced as a subset of cryo-EM methods, and many researchers may think 3D ED experiments can only be performed in cryogenic conditions. In this work, we demonstrated the structure determination of sucrose and H_4ABTC using 3D ED datasets. Furthermore, we showed the influence of temperature on data quality. In addition, we investigated the effects of data merging on structure determination under both temperature conditions. The understanding of the effect of temperature and data merging will help researchers to decide which temperature condition they should use during 3D ED data collection.

2. Materials and Methods

We chose two low-symmetry crystals, sucrose and azobenzene tetracarboxylic acid (H_4ABTC, 358.3 Da), as testing samples for the temperature comparison because these structures are stable under high vacuum in both temperature conditions. The sucrose crystals were obtained from a sugar cube for coffee and tea. The H_4ABTC molecules were synthesized according to the literature [32] and H_4ABTC crystals were formed from the unreacted linker during synthesis of a Zr-based metal–organic framework (MOF) [33]. The samples were crushed into fine powders in a mortar and deposited onto a lacey carbon TEM grid without using any solvents. 3D ED data were collected on a JEOL JEM-2100 with a LaB_6 filament at both room temperature and cryogenic temperature. Two grids were prepared for each sample. These grids were loaded on a Gatan 914 cryo-transfer holder and half of them were cooled down from ~300 K to ~100 K in the microscope column. Data were collected by continuously rotating the crystal while collecting selected area electron diffraction (SAED) patterns simultaneously with an electron-diffraction-dedicated hybrid pixel detector (Timepix, Amsterdam Scientific Instrument, Amsterdam, The Netherlands) running in continuous exposure mode [10,23]. Data were processed and merged by XDS [34]. The structures could be solved by using both SHELXT [35] and SIR2014 [36]. Structure refinement was performed using SHELXL [37,38].

3. Results and Discussion

3.1. Sucrose

For room and cryogenic temperature data collection, the crystals were continuously rotated at a constant speed of 1.13°/s. By using an exposure time of 0.3 s/frame, approximately 180 electron diffraction patterns were collected in each dataset, covering a total rotation angle of 60° on average. The data collection time on each crystal was less than 1 min, with an accumulated electron dose of less than 5 $e^-/Å^2$ (dose rate: 0.08 $e^-/Å^2/s$). In order to achieve high data completeness and redundancy, a total of 17 datasets were processed and merged using XDS (refer to Table 1). The resolution of these electron diffraction data extended up to 0.8 Å. All 23 non-H atoms in the asymmetric unit were located directly by using SHELXT, as well as SIR2014 (Figure 1C). The atom assignment was corrected manually since it is difficult for the software to distinguish C and O atoms with diffraction data. We first refined the structure without any hydrogen atoms and restraints for fair comparison. During structure refinement, the atomic displacement parameter (ADP) of more than half of the atoms was negative for datasets collected at liquid nitrogen temperature, as shown in Table 1 (refinement for comparison). In order to obtain the hydrogen position and compare the structure with a reference model, we then applied restraints in refinements against both datasets to get chemically sensible structures.

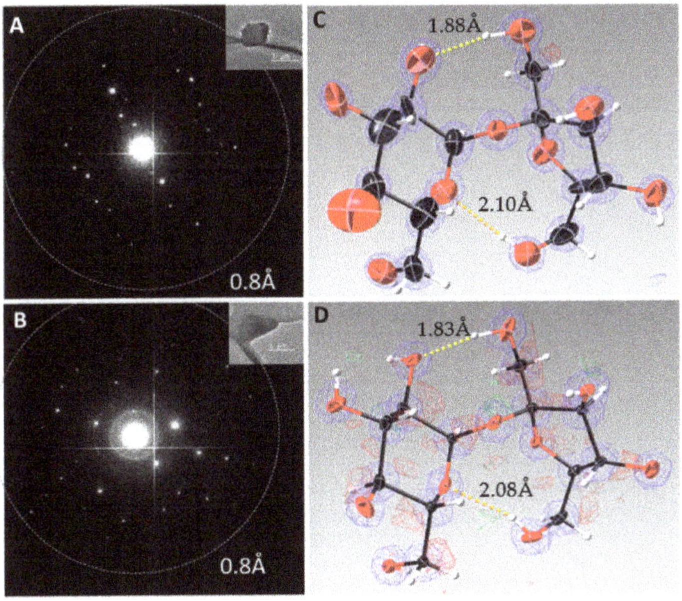

Figure 1. A typical electron diffraction pattern of sucrose crystal collected at (**A**) room temperature, (**B**) cryogenic condition. Note that the ice rings are visible in (**B**). The insets in (**A**,**B**) are typical crystals selected for data collection. Refined structure model of sucrose (asymmetric unit) against merged data collected at (**C**) room temperature, (**D**) cryogenic condition. The potential maps were obtained from SHELXL and visualized by SHELXLE. Blue volume means observed potential. Red volume means negative difference potential and green volume means positive difference potential. The threshold of observed and difference potential was set to 2σ (0.65 V) and 4σ (0.2 V), respectively.

Table 1. Data and refinement statistics and structure parameters from merged sucrose datasets.

	~300 K (0.80 Å)	~100 K (0.90 Å)
	Data Processing	
Number of crystals	17	12
Space group	$P2_1$	$P2_1$
a (Å)	11.19 [7]	10.91 [10]
b (Å)	9.06 [7]	8.78 [9]
c (Å)	8.04 [12]	7.85 [9]
β (°)	103.6 [10]	103.4 [8]
Resolution (Å)	0.8	0.9
$I/\sigma(I)$	3.61 (1.32)	3.31 (2.24)
$CC_{1/2}$ (%)	90.7 * (32.0 *)	78.9 * (50.1 *)
Redundancy	16.60	16.32
R_{meas}	0.619 (0.897)	0.617 (0.830)
No. of reflections	28601	18496
No. of unique reflections	3203	2093
Completeness	0.995 (0.964)	0.999 (1)
	Refinement (for comparison)	
No. of parameters	93 (isotropic)	93 (isotropic)
No. of restraints	1	1
R_1 ($Fo > 4\sigma(Fo)$)	0.2024	0.2330
R_1 all	0.2225	0.2401
GooF	1.102	1.063
wR_2	0.4974	0.4978
Number of negative ADPs	0	14
	Refinement (for obtaining better structure)	
No. of parameters	244 (anisotropic)	243 (anisotropic)
No. of restraints	5	4
R_1 ($F_o > 4\sigma(F_o)$)	0.1456	0.1974
R_1 all	0.1677	0.2058
R_{int}	0.5544	0.5205
GooF	1.021	0.945
wR_2	0.3981	0.4085
Hydrogen atoms found	40	40
Chemical formula	$C_{24}H_{40}O_{22}$	$C_{24}H_{40}O_{22}$
Weight (Da)	680.56	680.56

Values in brackets are the data statistics of the highest resolution shell. The resolution cut was decided by the statistical significance, which is denoted by an asterisk beside the $CC_{1/2}$ value. The square brackets behind the unit cell parameters stand for standard deviations. The unit cell parameters from the reference model were used in the refinements against both datasets.

To better compare the structure refinements against datasets obtained at room temperature and cryogenic temperature, we performed structural analyses in two rounds. In the first round, we used the same SHELX input file as the starting point and refined the model against both datasets for comparison. In the second round, we further refined the structure in order to obtain the best structure models achievable from the two datasets.

In the first round, we refined the structure against both datasets isotropically using the same number of parameters (93) with one restraint. As shown in Table 1, the final R_1 value is 0.2330 for data collected at cryogenic temperature and 0.2024 for data collected at room temperature. Furthermore, the structure obtained from the cryogenic temperature dataset contained 14 atoms with negative ADPs, while all the ADPs were positive in the structure obtained from the room temperature dataset.

In the second round, we further refined the structures against both datasets with anisotropic refinement and "XNPD" keyword to obtain the best structures. The "XNPD" keyword sets a lower bound for the eigenvalues of the U_{ij} tensor of all anisotropic atoms or the U of an isotropic atom. For the room temperature dataset, a total of 244 parameters were refined using five restraints. Four of these restraints are distance restraints between carbon atoms and hydrogen atoms, and one of them is an ADP restraint. The final

R_1 value is 0.1456 for 2147 strong reflections with $F_o > 4\sigma(F_o)$, and 0.1677 for all 3203 reflections. The atomic co-ordinates of the refined structure were compared with those of sucrose structures deposited in the CCDC database, which are obtained from neutron diffraction [39]. Since we are interested in investigating the feasibility of refining hydrogen atoms, the sucrose models were compared with this reference structure [40]. The average deviations from the reference model (ADRAs) are 0.05(2) Å for C and O atoms, and 0.17(9) Å for H atoms (Table 2 and supplementary material Table S1). The average bond lengths after refinement are 1.51(3) Å for C–C bonds, and 1.40(2) Å for C–O bonds (Table 3). The intermolecular and intramolecular hydrogen bond network could also be resolved by using the merged dataset collected at room temperature (Figures 1C and 2). These refinement results showed that, although the R_{int} value is quite high, accurate structure was obtained after the refinement, possibly due to the increase in completeness and data redundancy. The significant increase in R_{int} value has two contributing factors. The first being that the dynamical effects are ignored during data merging. Crystals of different size, shape, and orientation on the TEM grids contain different dynamical effects. However, current data-merging software developed for X-ray crystallography does not take this into account. The scaling of the intensities is far from optimal. The second factor is radiation damage. Even though the electron dose rate used in this study was very low (0.08 $e^-/Å^2/s$), we can only collect datasets with an average tilt range of 60°, since sucrose crystal is very beam sensitive. Even though data merging software has modelled radiation damage (e.g., DECAY in XSCALE), merging different datasets still accumulates errors and inconsistencies into the final merged HKL file, leading to a large R_{int} value.

The statistics of the data collected at a cryogenic temperature can be found in Table 1. By comparing the data quality and difference Fourier map with data collected at room temperature, it is clear that the data quality is higher and difference Fourier map is cleaner for data collected at room temperature. Although the refinement statistics from 100 K data are comparable to those refined using data collected at room temperature, the atomic displacement parameters were constrained by using XNPD keyword. Otherwise, 14 atoms would have negative ADPs, as shown in Table 1. On the other hand, data collected at 100 K, in general, have lower overall $I/\sigma(I)$ (3.31) compared with the overall $I/\sigma(I)$ from room temperature data (3.61) due to the ice contamination and inelastic scattering, as shown in Figure S1. The crystal mosaicity might increase during the cooling process and the reflection peaks would become less "sharp". The refined structured model has an average ADRA of 0.06(2) for C and O atoms (Table 2) and 0.20(11) for H atoms (Table S1). The average bond lengths are 1.54(2) Å for C–C bonds and 1.39(3) Å for C–O bonds (Table 3). The ADPs from the refinement with XNPD keyword showed that the ADP values for some atoms (C1, C9) reached 0.01, which is the minimum value imposed by XNPD keyword (Table S2). For experiments using a cryo-transfer holder and a microscope without a dedicated anticontamination device, it is rather difficult to avoid ice contaminations. The inelastic scattering events are increased due to microscopic ice crystals forming in the background. "Ice rings" are found in the diffraction pattern (Figure 1B). Their intensities increase towards the end of data collection because of accumulated contamination. In some cases, cryogenic conditions can even cause the deformations of crystals. In addition, we compared the difference Fourier maps in Figure 1C,D, and the map from merged room temperature datasets is much cleaner than other maps. Therefore, if the sample is stable under vacuum conditions and stable under electron beam, it is generally advisable to collect 3D ED data under ambient conditions.

Table 2. Deviations of atomic positions between the reference structure of sucrose [39] and the structure determined from 3D ED datasets under both temperature conditions. The average deviations are 0.05(2) Å for C and O atoms for the merged dataset and 0.06(2) Å for the merged dataset collected at low temperature.

Atom Label	ADRA (Å)	Atom Label	ADRA (Å)
~300 K		~100 K	
C1	0.027	C1	0.077
C2	0.076	C2	0.046
C3	0.022	C3	0.056
C4	0.029	C4	0.064
C5	0.044	C5	0.049
C6	0.046	C6	0.038
C7	0.064	C7	0.021
C8	0.071	C8	0.073
C9	0.047	C9	0.070
C10	0.060	C10	0.059
C11	0.065	C11	0.064
C12	0.063	C12	0.082
O1	0.035	O1	0.045
O2	0.041	O2	0.078
O3	0.079	O3	0.055
O4	0.057	O4	0.058
O5	0.092	O5	0.093
O6	0.033	O6	0.039
O7	0.090	O7	0.059
O8	0.059	O8	0.070
O9	0.020	O9	0.024
O10	0.031	O10	0.046
O11	0.036	O11	0.021
Average	**0.05(2)**	**Average**	**0.06(2)**

Table 3. Bond lengths found in the crystal structure of sucrose, which were refined from electron diffraction data.

Atom1	Atom2	Bond Length (Å)	Atom1	Atom2	Bond Length (Å)
~300 K			~100 K		
C1	C3	1.533	C1	C3	1.581
C2	C4	1.477	C2	C4	1.510
C3	C5	1.515	C3	C5	1.535
C4	C6	1.510	C4	C6	1.521
C5	C7	1.552	C5	C7	1.559
C6	C8	1.556	C6	C8	1.565
C7	C9	1.488	C7	C9	1.563
C8	C10	1.482	C8	C10	1.524
C9	C11	1.517	C9	C11	1.537
C10	C12	1.489	C10	C12	1.509

Table 3. *Cont.*

Atom1	Atom2	Bond Length (Å)	Atom1	Atom2	Bond Length (Å)
C–C Average		**1.51(3)**	C–C Average		**1.54(2)**
C1	O1	1.427	C1	O1	1.402
C1	O9	1.374	C1	O9	1.377
C2	O2	1.432	C2	O2	1.497
C3	O3	1.367	C3	O3	1.361
C4	O1	1.410	C4	O1	1.403
C4	O4	1.391	C4	O4	1.453
C5	O5	1.375	C5	O5	1.385
C6	O6	1.360	C6	O6	1.331
C7	O7	1.416	C7	O7	1.633
C8	O8	1.423	C8	O8	1.499
C9	O9	1.399	C9	O9	1.485
C10	O4	1.437	C10	O4	1.347
C11	O10	1.411	C11	O10	1.433
C12	O11	1.420	C12	O11	1.455
C–O Average		**1.40(2)**	C–O Average		**1.39(3)**

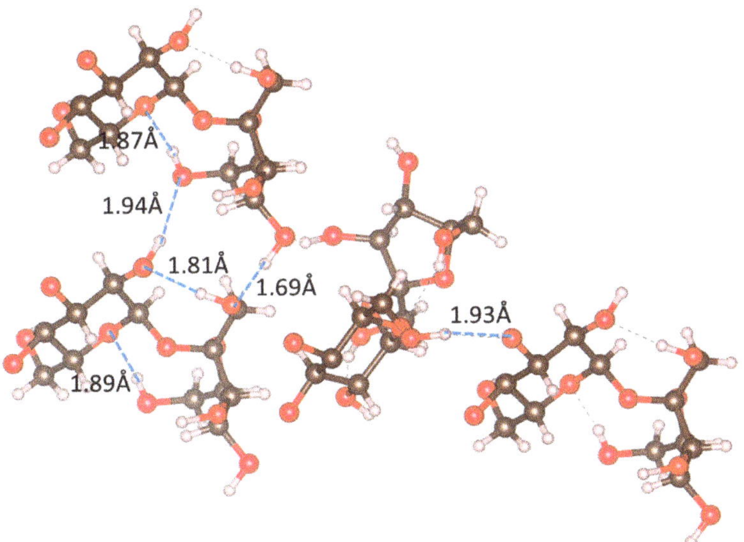

Figure 2. Details of the best sucrose structure obtained from merged room temperature data. The hydrogen bond network is highlighted.

3.2. H_4ABTC

To further extend our observations, we use H_4ABTC as another example. To our knowledge, no crystal structure of H_4ABTC molecule has been reported previously. 3D ED datasets of H_4ABTC crystals were collected at room temperature and under cryogenic conditions. The crystals were continuously rotated at a constant speed of 0.46°/s. By using an exposure time of 0.5 s/frame, approximately 435–480 frames are collected in each dataset, covering a total rotation angle of 100–110° on average. The data collection time on each crystal is approximately 4 min, with an accumulated electron dose of less than 20 $e^-/Å^2$ (dose rate: 0.08 $e^-/Å^2/s$). For datasets collected at room temperature, the resolution of these electron diffraction data extended up to 0.80 Å. We merged multiple datasets to achieve higher completeness and data redundancy for structure determination. Direct methods were applied for the structure solution and all 78 non-H atoms in the asymmetric unit could be directly located using either SHELXT or SIR2014. The structure was refined using data up to 0.83 Å resolution using SHELXL (Table 4). By examining the raw diffraction patterns (Figure 3A,B), we observed ice rings in all cryogenic temperature datasets. The $I/\sigma(I)$ and $CC_{1/2}$ in Table 4 indicates the data quality was reduced at high resolution in cryogenic datasets. H_4ABTC crystallizes in the space group P-1, with three molecules in an asymmetric unit. The packing of these three molecules is shown in Figure 4A,B. Two molecules were parallel and had strong π–π interactions between them, while the third one laid perpendicular to these two molecules. We adapted the evaluation strategy of sucrose crystals to H_4ABTC crystals. We first used isotropic refinement with 314 parameters and no restraints for both merged datasets for the purpose of comparison. No hydrogen atoms were added and no negative ADPs were found during refinement. The final R_1 obtained for the room temperature dataset and cryogen dataset was 0.2596 and 0.2833, respectively.

Table 4. Data and refinement statistics and structure parameters from merged H_4ABTC datasets.

	~300 K (0.8 Å)	**~100 K (0.9 Å)**
	Data Processing	
Number of crystals	6	9
Space group	P-1	P-1
a (Å)	13.47 [6]	13.52 [7]
b (Å)	14.63 [12]	14.63 [17]
c (Å)	14.65 [12]	14.70 [13]
α (°)	98.55	99.40
β (°)	105.90	106.06
γ (°)	112.92	113.99
Resolution (Å)	0.80	0.90
$I/\sigma(I)$	4.77 (1.99)	2.63 (1.35)
$CC_{1/2}$ (%)	97.1 * (88.4 *)	90.8 * (73.0 *)
Redundancy	4.89	7.08
R_{meas}	0.209 (0.595)	0.395 (0.648)
No. of reflections	48822	48984
No. of unique reflections	8936	6770
Completeness	0.897 (0.814)	0.935 (0.752)
	Refinement (for comparison)	
No. of parameters	314 (isotropic)	314 (isotropic)
No. of restraints	0	0
R_1 ($F_o > 4\sigma(F_o)$)	0.2596	0.2833
R_1 all	0.2878	0.3329
R_{int}	0.1843	0.3353
GooF	1.115	1.055
wR_2	0.6227	0.6858
	Refinement (for obtaining better structure)	
No. of parameters	719 (anisotropic)	703 (anisotropic)
No. of restraints	77	77
R_1 ($F_o > 4\sigma(F_o)$)	0.1883	0.2029
R_1 all	0.2146	0.2472
R_{int}	0.1843	0.3353
GooF	1.142	1.085
wR_2	0.4689	0.5273
Chemical formula	$C_{96}H_{52}N_{12}O_{48}$	$C_{96}H_{36}N_{12}O_{48}$
Weight	2141.51	2125.38
Hydrogen atoms found	52	36

Values in brackets are the data statistics of the highest resolution shell. The resolution cut was decided by the statistical significance, which is denoted by an asterisk beside the $CC_{1/2}$ value. The square brackets behind the unit cell parameters stand for standard deviations. The unit cell parameters from the reference model were used in the refinements against both datasets.

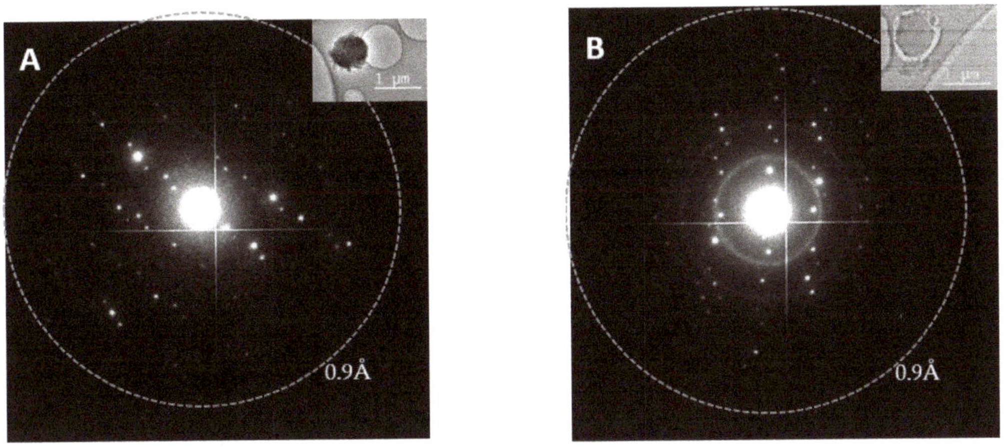

Figure 3. (**A**) Typical electron diffraction pattern of H_4ABTC crystal collected at room temperature. (**B**) Typical electron diffraction pattern of H_4ABTC crystal collected under cryogenic conditions. The insets are images of typical microcrystals.

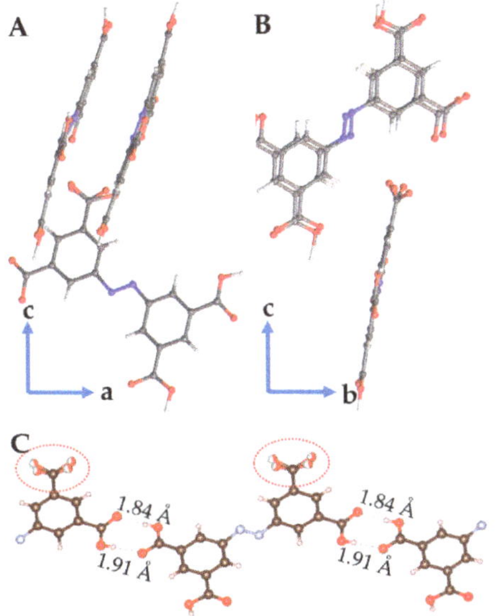

Figure 4. Detailed structure of the H_4ABTC molecule refined from electron diffraction data collected under room temperature conditions, viewed along the (**A**) *b*-axis and (**B**) *a*-axis. (**C**) The terminal carboxyl groups are more flexible. The oxygen atoms in these groups are refined as disorders.

In order to obtain the best structures from both datasets, we further refined the ADPs anisotropically and only added the hydrogen atoms according to the difference Fourier map. "ISOR" restraints were used to avoid negative ADPs during refinement. The room temperature dataset was refined using 719 parameters with 77 restraints. The final R_1 value is 0.1883 for 6176 strong reflections with $F_o > 4\sigma(F_o)$ and 0.2146 for all 8143 reflections. The bond lengths are analyzed and shown in Table 5. The average bond lengths are 1.52(2) Å for single C–C bonds, 1.43(2) Å for C–C bonds in a benzene ring, 1.30(4) Å for C–O bonds, 1.27(4) Å for C=O bonds, 1.45(2) Å for C–N bonds, and 1.27(1) Å for N=N bonds. It is possible to distinguish some hydroxyl groups and carbonyl groups by analyzing the bond lengths between C and O in the refined model. Furthermore, we could refine the disorder of the oxygen atoms in the terminal carboxyl groups with electron diffraction data (Figure 4C). Out of the 60 hydrogen atoms, 52 hydrogen positions, especially those associated with benzene rings, can be located from q-peaks, as well as from the difference Fourier map.

Table 5. Mean and standard deviation (in brackets) of bond length of H_4ABTC structure solved from 3D ED datasets collected at room temperature and cryogenic temperature.

Bond Type	Bond Length (Å) (~300 K)	Bond Length (Å) (~100 K)
C–C single bond	1.52 (2)	1.53 (2)
C–C benzene	1.43 (2)	1.43 (3)
C–O carboxyl	1.30 (4)	1.34 (7)
C=O carboxyl	1.27 (4)	1.29 (6)
C–N single bond	1.45 (2)	1.46 (3)
N=N double bond	1.27 (1)	1.26 (4)

Similar to the sucrose case, datasets collected at 100 K produced worse refinement statistics (Table 4). The resolution is lower (0.8 Å to 0.9 Å) and the overall $I/\sigma(I)$ is much lower (4.77 to 2.63) due to the higher background estimation by XDS (Figure S2). Subsequently, the refinement results are worse than the results from room temperature datasets. The final R_1 reached 0.2029 for strong reflections and 0.2472 for all reflections. The structure was refined with 703 parameters together with 77 restraints. We applied the same "ISOR" restraints as the room temperature dataset. We compared the anisotropic ADPs from both datasets and we found that the ADPs obtained from room temperature datasets are less elongated compared with ADPs from cryogenic temperature datasets, as shown in Figure 5. We think preferred orientation contributed to the elongation of the ADP ellipsoids. When a plate-like crystal is tilted to a high angle, the thickness of the sample increases, while the number of unit cells perpendicular to the electron beam decreases, causing the intensities of the reflections to decrease. The 3D ED data become anisotropic for plate-like crystals. This could become even more severe for cryogenic temperature measurement because of increased background. Next, we calculated the average bond lengths and the standard deviation for different types of bonds. Although the average bond length is almost the same as the bond length from the room temperature dataset, the standard deviation is greater, as shown in Table 5. In total, 36 hydrogen atoms could be located in the difference Fourier maps from the cryogenic temperature dataset. We note that the difference in the number of refined parameters (room temperature vs. cryogenic) is due to the fact that fewer hydrogen atoms were found from the cryogenic temperature dataset. Moreover, in both structure refinement studies of H_4ABTC, all the hydrogen atoms located on carboxyl groups were lost. The location of hydrogen atoms can be classified into three groups: (a) located on the benzene ring, (b) located between two carboxyl groups with hydrogen bonding interaction, and (c) carboxyl groups with no interactions. From our experience, the hydrogen atoms located on the benzene ring are the easiest to find. Hydrogen atoms in group (c) are the most difficult to locate because of the rotation of carboxyl groups, which also leads to the observed disorder in Figure 4C. We also noticed that the disordered oxygen atoms in the terminal carboxyl group became distorted for datasets collected under cryogenic conditions, while, for the room temperature data, the geometry of the disordered carboxyl group is consistent with prior chemical knowledge. These results indicate that the data quality and structural details obtained from datasets collected at room temperature are comparable with those obtained from cryogenic temperature datasets.

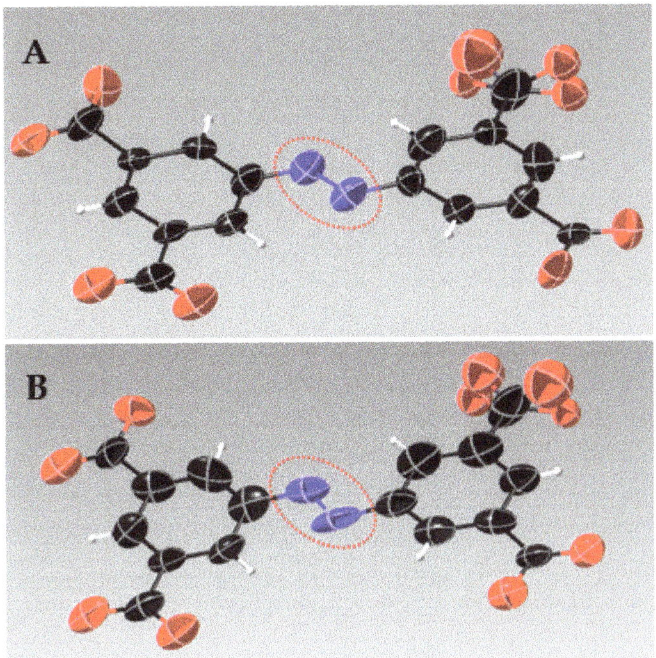

Figure 5. Structure comparison of one H_4ABTC molecule from refinement for obtaining the best structures at (**A**) room temperature and at (**B**) cryogenic temperature.

4. Conclusions

In this work, we determined the structure of two small organic molecules by 3D ED in different temperature conditions. Among these two molecules, the structure of H_4ABTC is unknown and it is the first crystal structure ever reported for this molecule. Based on the comparisons made between 3D ED datasets of sucrose and H_4ABTC collected at room temperature and cryogenic conditions, we conclude that, if crystals are stable under vacuum and electron beam, comparable or better data quality and structure accuracy can be achieved by collecting data at room temperature. The fine details, such as hydrogen positions, disorders, and ADPs, are better revealed from merged room temperature datasets. However, under cryogenic conditions, the resolution and $I/\sigma(I)$ ratio could be lower if a dedicated cryo-electron microscope is not available. If the crystal is stable under vacuum conditions in TEM and electron beam, it is better to collect 3D ED data at room temperature, which is also simpler and more accessible. Otherwise, cryo-transfer and cooling techniques can be utilized. 3D ED is a capable method for structure determination of small organic molecules from nano- and micron-sized crystals, whether measured at liquid nitrogen temperature or at room temperature.

Supplementary Materials: The following are available online at https://www.mdpi.com/2073-8994/13/11/2131/s1, Figure S1: Estimated background by XDS of sucrose data collected at (A) room temperature and (B) cryogenic temperature, Figure S2: Estimated background by XDS of H_4ABTC data collected at (A) room temperature and (B) cryogenic temperature, Table S1: ADRA of hydrogen atoms between the reference structure of sucrose [1] and structures determined from datasets obtained in both temperature conditions (2nd round of refinements for obtaining the best possible structure models), Table S2: Comparison between the ADPs of sucrose determined from merged datasets (refinement for obtaining better structure). ADPs close to 0.01 (the minimum ADP value set by "XNPD") are marked in red.

Author Contributions: Conceptualization, T.Y., X.Z. and H.X.; methodology, all authors; validation, T.Y., X.Z. and H.X.; formal analysis, T.Y. and H.X.; resources, N.S., X.Z. and H.X.; data curation, H.X.; writing-original draft preparation, T.Y. and H.X.; writing-review and editing, T.Y., X.Z., H.X., N.S., A.K.I. and S.W.; visualization, T.Y. and H.X.; supervision, N.S., X.Z. and H.X.; funding acquisition, X.Z. and H.X. All authors have read and agreed to the published version of the manuscript.

Funding: The project is supported by the Swedish Research Council (2017-05333, H.X.; 2019-00815, X.Z.), the Knut and Alice Wallenberg Foundation (2018.0237, X.Z.), and the Science for Life Laboratory through the technique development grant (MicroED@SciLifeLab).

Institutional Review Board Statement: Not applicable.

Informed Consent Statement: Not applicable.

Data Availability Statement: The CIF files of the structures are deposited to the Cambridge Structure Database (CSD) with deposition codes: 2112205, 2112206, 2112207, 2112208, 2112209, 2112210, 2112211, 2112212.

Conflicts of Interest: The authors declare no conflict of interest.

References

1. Clabbers, M.T.B.; Abrahams, J.P. Electron diffraction and three-dimensional crystallography for structural biology. *Crystallogr. Rev.* **2018**, *24*, 176–204. [CrossRef]
2. Dorset, D.L.; Hauptman, H.A. Direct phase determination for quasi-kinematical electron diffraction intensity data from organic microcrystals. *Ultramicroscopy* **1976**, *1*, 195–201. [CrossRef]
3. Dorset, D.L. *Structural Electron Crystallography;* Springer: Boston, MA, USA, 1995; ISBN 978-0-306-45049-5.
4. Dong, W.; Baird, T.; Fryer, J.R.; Gilmore, C.J.; MacNicol, D.D.; Bricogne, G.; Smith, D.J.; O'Keefe, M.A.; Hovmöller, S. Electron microscopy at 1-Å resolution by entropy maximization and likelihood ranking. *Nature* **1992**, *355*, 605–609. [CrossRef]
5. Kolb, U.; Gorelik, T.; Kübel, C.; Otten, M.; Hubert, D. Towards automated diffraction tomography: Part I—Data acquisition. *Ultramicroscopy* **2007**, *107*, 507–513. [CrossRef] [PubMed]
6. Kolb, U.; Gorelik, T.; Otten, M. Towards automated diffraction tomography. Part II—Cell parameter determination. *Ultramicroscopy* **2008**, *108*, 763–772. [CrossRef] [PubMed]
7. Zhang, D.; Oleynikov, P.; Hovmöller, S.; Zou, X. Collecting 3D electron diffraction data by the rotation method. *Z. Für Krist. Int. J. Struct. Phys. Chem. Asp. Cryst. Mater.* **2010**, *225*, 94–102. [CrossRef]
8. Wan, W.; Sun, J.; Su, J.; Hovmöller, S.; Zou, X. Three-dimensional rotation electron diffraction: SoftwareREDfor automated data collection and data processing. *J. Appl. Crystallogr.* **2013**, *46*, 1863–1873. [CrossRef]
9. Kolb, U.; Gorelik, T.E.; Mugnaioli, E.; Stewart, A. Structural Characterization of Organics Using Manual and Automated Electron Diffraction. *Polym. Rev.* **2010**, *50*, 385–409. [CrossRef]
10. Van Genderen, E.; Clabbers, M.T.B.; Das, P.P.; Stewart, A.; Nederlof, I.; Barentsen, K.C.; Portillo, Q.; Pannu, N.S.; Nicolopoulos, S.; Gruene, T.; et al. Ab initiostructure determination of nanocrystals of organic pharmaceutical compounds by electron diffraction at room temperature using a Timepix quantum area direct electron detector. *Acta Crystallogr. Sect. A Found. Adv.* **2016**, *72*, 236–242. [CrossRef]
11. Gruene, T.; Wennmacher, J.T.C.; Zaubitzer, C.; Holstein, J.J.; Heidler, J.; Fecteau-Lefebvre, A.; De Carlo, S.; Müller, E.; Goldie, K.N.; Regeni, I.; et al. Rapid Structure Determination of Microcrystalline Molecular Compounds Using Electron Diffraction. *Angew. Chem. Int. Ed.* **2018**, *57*, 16313–16317. [CrossRef] [PubMed]
12. Jones, C.G.; Martynowycz, M.W.; Hattne, J.; Fulton, T.J.; Stoltz, B.M.; Rodriguez, J.A.; Nelson, H.M.; Gonen, T. The CryoEM Method MicroED as a Powerful Tool for Small Molecule Structure Determination. *ACS Cent. Sci.* **2018**, *4*, 1587–1592. [CrossRef]
13. Clabbers, M.T.B.; Gruene, T.; Van Genderen, E.; Abrahams, J.P. Reducing dynamical electron scattering reveals hydrogen atoms. *Acta Crystallogr. Sect. A Found. Adv.* **2019**, *75*, 82–93. [CrossRef] [PubMed]
14. Jones, C.G.; Asay, M.; Kim, L.J.; Kleinsasser, J.F.; Saha, A.; Fulton, T.J.; Berkley, K.R.; Cascio, D.; Malyutin, A.; Conley, M.P.; et al. Characterization of Reactive Organometallic Species via MicroED. *ACS Central Sci.* **2019**, *5*, 1507–1513. [CrossRef] [PubMed]
15. Das, P.P.; Andrusenko, I.; Mugnaioli, E.; Kaduk, J.A.; Nicolopoulos, S.; Gemmi, M.; Boaz, N.C.; Gindhart, A.M.; Blanton, T. Crystal Structure of Linagliptin Hemihydrate Hemiethanolate $(C_{25}H_{28}N_8O_2)_2(H_2O)(C_2H_5OH)$ from 3D Electron Diffraction Data, Rietveld Refinement, and Density Functional Theory Optimization. *Cryst. Growth Des.* **2021**, *21*, 2019–2027. [CrossRef]
16. Halaby, S.; Martynowycz, M.W.; Zhu, Z.; Tretiak, S.; Zhugayevych, A.; Gonen, T.; Seifrid, M. Microcrystal Electron Diffraction for Molecular Design of Functional Non-Fullerene Acceptor Structures. *Chem. Mater.* **2021**, *33*, 966–977. [CrossRef]
17. Gleason, P.R.; Nannenga, B.L.; Mills, J.H. Rapid Structural Analysis of a Synthetic Non-canonical Amino Acid by Microcrystal Electron Diffraction. *Front. Mol. Biosci.* **2021**, *7*, 461. [CrossRef]
18. Palatinus, L.; Brázda, P.; Boullay, P.; Perez, O.; Klementová, M.; Petit, S.; Eigner, V.; Zaarour, M.; Mintova, S. Hydrogen positions in single nanocrystals revealed by electron diffraction. *Science* **2017**, *355*, 166–169. [CrossRef]

19. Brázda, P.; Palatinus, L.; Babor, M. Electron diffraction determines molecular absolute configuration in a pharmaceutical nanocrystal. *Science* **2019**, *364*, 667–669. [CrossRef]
20. Shi, D.; Nannenga, B.; Iadanza, M.; Gonen, T. Three-dimensional electron crystallography of protein microcrystals. *eLife* **2013**, *2*, e01345. [CrossRef] [PubMed]
21. Nannenga, B.; Shi, D.; Leslie, A.G.W.; Gonen, T. High-resolution structure determination by continuous-rotation data collection in MicroED. *Nat. Methods* **2014**, *11*, 927–930. [CrossRef]
22. Yonekura, K.; Kato, K.; Ogasawara, M.; Tomita, M.; Toyoshima, C. Electron crystallography of ultrathin 3D protein crystals: Atomic model with charges. *Proc. Natl. Acad. Sci. USA* **2015**, *112*, 3368–3373. [CrossRef]
23. Nederlof, I.; Van Genderen, E.; Li, Y.-W.; Abrahams, J.P. A Medipix quantum area detector allows rotation electron diffraction data collection from submicrometre three-dimensional protein crystals. *Acta Crystallogr. Sect. D Biol. Crystallogr.* **2013**, *69*, 1223–1230. [CrossRef] [PubMed]
24. Clabbers, M.T.B.; Van Genderen, E.; Wan, W.; Wiegers, E.L.; Gruene, T.; Abrahams, J.P. Protein structure determination by electron diffraction using a single three-dimensional nanocrystal. *Acta Crystallogr. Sect. D Struct. Biol.* **2017**, *73*, 738–748. [CrossRef]
25. Lanza, A.; Margheritis, E.; Mugnaioli, E.; Cappello, V.; Garau, G.; Gemmi, M. Nanobeam precession-assisted 3D electron diffraction reveals a new polymorph of hen egg-white lysozyme. *IUCrJ* **2019**, *6*, 178–188. [CrossRef]
26. Gemmi, M.; Lanza, A.E. 3D electron diffraction techniques. *Acta Crystallogr. Sect. B Struct. Sci. Cryst. Eng. Mater.* **2019**, *75*, 495–504. [CrossRef] [PubMed]
27. Gallagher-Jones, M.; Bustillo, K.C.; Ophus, C.; Richards, L.S.; Ciston, J.; Lee, S.; Minor, A.M.; Rodriguez, J.A. Atomic structures determined from digitally defined nanocrystalline regions. *IUCrJ* **2020**, *7*, 490–499. [CrossRef]
28. Xu, H.; Lebrette, H.; Yang, T.; Srinivas, V.; Hovmöller, S.; Högbom, M.; Zou, X. A Rare Lysozyme Crystal Form Solved Using Highly Redundant Multiple Electron Diffraction Datasets from Micron-Sized Crystals. *Structure* **2018**, *26*, 667–675.e3. [CrossRef] [PubMed]
29. Xu, H.; Lebrette, H.; Clabbers, M.T.B.; Zhao, J.; Griese, J.J.; Zou, X.; Högbom, M. Solving a new R2lox protein structure by microcrystal electron diffraction. *Sci. Adv.* **2019**, *5*, eaax4621. [CrossRef] [PubMed]
30. Clabbers, M.T.B.; Holmes, S.; Muusse, T.W.; Vajjhala, P.R.; Thygesen, S.J.; Malde, A.K.; Hunter, D.J.B.; Croll, T.I.; Flueckiger, L.; Nanson, J.D.; et al. MyD88 TIR domain higher-order assembly interactions revealed by microcrystal electron diffraction and serial femtosecond crystallography. *Nat. Commun.* **2021**, *12*, 2578. [CrossRef]
31. Clabbers, M.T.B.; Fisher, S.Z.; Coinçon, M.; Zou, X.; Xu, H. Visualizing drug binding interactions using microcrystal electron diffraction. *Commun. Biol.* **2020**, *3*, 1–8. [CrossRef]
32. Ameerunisha, S.; Zacharias, P.S. Characterization of simple photoresponsive systems and their applications to metal ion transport. *J. Chem. Soc. Perkin Trans. 2* **1995**, 1679–1682. [CrossRef]
33. Van Velthoven, N.; Waitschat, S.; Chavan, S.M.; Liu, P.; Smolders, S.; Vercammen, J.; Bueken, B.; Bals, S.; Lillerud, K.P.; Stock, N.; et al. Single-site metal–organic framework catalysts for the oxidative coupling of arenes via C–H/C–H activation. *Chem. Sci.* **2019**, *10*, 3616–3622. [CrossRef] [PubMed]
34. Gemmi, M.; La Placa, M.G.I.; Galanis, A.S.; Rauch, E.; Nicolopoulos, S. Fast electron diffraction tomography. *J. Appl. Crystallogr.* **2015**, *48*, 718–727. [CrossRef]
35. Kabsch, W. XDS. *Acta Crystallogr. Sect. D Biol. Crystallogr.* **2010**, *66*, 125–132. [CrossRef] [PubMed]
36. Sheldrick, G.M. SHELXT–Integrated space-group and crystal-structure determination. *Acta Crystallogr. Sect. A Found. Adv.* **2015**, *71*, 3–8. [CrossRef] [PubMed]
37. Burla, M.C.; Caliandro, R.; Carrozzini, B.; Cascarano, G.L.; Cuocci, C.; Giacovazzo, C.; Mallamo, M.; Mazzone, A.; Polidori, G. Crystal structure determination and refinementviaSIR2014. *J. Appl. Crystallogr.* **2015**, *48*, 306–309. [CrossRef]
38. Sheldrick, G.M. Crystal structure refinement withSHELXL. *Acta Crystallogr. Sect. C Struct. Chem.* **2015**, *71*, 3–8. [CrossRef]
39. Hübschle, C.B.; Sheldrick, G.M.; Dittrich, B. ShelXle: A Qt graphical user interface forSHELXL. *J. Appl. Crystallogr.* **2011**, *44*, 1281–1284. [CrossRef] [PubMed]
40. Brown, G.M.; Levy, H.A. Further refinement of the structure of sucrose based on neutron-diffraction data. *Acta Crystallogr. Sect. B Struct. Crystallogr. Cryst. Chem.* **1973**, *29*, 790–797. [CrossRef]

symmetry MDPI

Article

Identification of Retained Austenite in 9Cr-1.4W-0.06Ta-0.12C Reduced Activation Ferritic Martensitic Steel

Rengachari Mythili [1,2,*], Ravi Kirana [3], Loushambam Herojit Singh [4], Ramanujam Govindaraj [2,5], Anil K. Sinha [6], Manvendra N. Singh [7], Saibaba Saroja [1,†], Muraleedharan Vijayalakshmi [1,†] and Sudip K. Deb [7]

1 Metallurgy & Materials Group, Indira Gandhi Centre for Atomic Research, Kalpakkam 603102, India; saroja.saibaba@gmail.com (S.S.); vijayalakshmi.muraleedharan@gmail.com (M.V.)
2 Homi Bhabha National Institute at IGCAR, Kalpakkam 603102, India; govind@igcar.gov.in
3 Department of Physics and Nanotechnology, SRM Institute of Science and Technology, Kattankulathur 603203, India; ravikirs@srmist.edu.in
4 Department of Physics, NIT Manipur, Langol 795004, India; herojit@nitmanipur.ac.in
5 Materials Science Group, Indira Gandhi Centre for Atomic Research, Kalpakkam 603102, India
6 Department of Physics, University of Petroleum and Energy Studies, Dehradun 248007, India; anil.sinha@ddn.upes.ac.in
7 Raja Ramanna Centre for Advanced Technology, Indore 452013, India; manavendra@rrcat.gov.in (M.N.S.); skdeb@rrcat.gov.in (S.K.D.)
* Correspondence: rm@igcar.gov.in
† These authors have superannuated.

Abstract: Reduced activation ferritic martensitic (RAFM) 9Cr steels, which are candidate materials for the test blanket module (TBM) of nuclear fusion reactors, are considered to be air hardenable. However, alloy composition and the processing conditions play a significant role during the transformation of austenite to martensite/ferrite on cooling. The presence of retained austenite is known to influence the mechanical properties of the steel. Identifying very low amounts of retained austenite is very challenging though conventional microscopy. This paper aims at identifying a low amount of retained austenite in normalized 9Cr-1.4W-0.06Ta-0.12C RAFM steel using synchrotron X-ray diffraction and Mossbauer spectroscopy and confirmed by advanced automated crystal orientation mapping in transmission electron microscopy. Homogeneity of austenite has been understood to influence the microstructure of the normalized steel, which is discussed in detail.

Keywords: reduced activation ferritic martensitic steel; microstructure; martensite; retained austenite; automated crystal orientation mapping-transmission electron microscopy; Mossbauer spectroscopy

1. Introduction

Reduced activation ferritic martensitic (RAFM) 9Cr steels are considered to be candidate materials for test blanket module (TBM) for nuclear fusion reactors, due to their excellent mechanical properties [1–3]. They are also a very good structural materials for steam generator circuits [4,5] as they exhibit good thermal conductivity and low thermal expansion coefficient [6,7].

The 9Cr ferritic/martensitic steels are reported to be air hardenable [8], after normalization above Ac_3 temperature. These steels with a low carbon content (<0.15%) are known to form a lath martensitic structure with a dislocation substructure [9,10]. Though martensite gives good strength to the steel, due to limited ductility, these steels are not recommended to be used in normalized condition [11,12]. Therefore, the steels are generally used in the normalized and tempered condition with a tempered martensitic structure possessing good mechanical properties. However, preceding the tempering treatment, normalizing the steel in the single austenite phase is also an essential step for the transformation to a complete martensitic structure [11–13]. The decomposition of austenite is known to be influenced by the rate of cooling (Q) from the austenitizing temperature and the alloy chemistry [11,12].

Continuous cooling transformation (CCT) and time temperature transformation (TTT) of 9 Cr steels have been already studied [10,14]. As per the existing literature, CCT or TTT diagrams explain the presence of martensite/ferrite and carbides depending on the cooling rates adopted, but not the presence of retained austenite in 9Cr ferritic steels [9–12]. Martensitic transformation of austenite is well known to be accompanied by strain dictated by the composition of the steel and the cooling rates and hence the transformation induced strain can induce the mechanical stabilization of the austenite, and the mechanical stability of retained austenite is important in obtaining good toughness [15]. In fact, residual stress introduced unintentionally in weldments during fabrication can augment the transformation strain that could impede complete martensitic transformation of austenite. It is reported that mechanical stabilization might play an even greater role for retained austenite in the weldments of modified 9Cr–1Mo steel, which exhibits higher creep strength than most low alloy steels [15]. Santella et al. [16] report an incomplete austenite transformation in the weld metal of modified 9Cr-1Mo steel during initial cooling from an austenitizing treatment and the transformation of retained austenite to martensite during tempering, and attributed to the effects of microsegregation. The presence of retained austenite can be undesirable in the steel, if it is metastable, and can lead to the formation of untempered martensite or inhomogeneity after tempering [17]. Inhomogeneity in such steel components can have a significant negative impact on the mechanical properties, reducing the service life. During tempering, decomposition of retained austenite to different carbides is influenced by tempering temperature and time, which in turn, are responsible for temper embrittlement of the steel [18,19]. Therefore, it is pertinent to identify the reasons for the presence of retained austenite in such steels and also characterize using appropriate techniques, since the amount of retained austenite could be very low. Hence, in this study, a detailed microstructural characterization of RAFM steel in normalized as well as tempered condition has been taken up, to understand the decomposition mode of high temperature austenite.

2. Materials and Methods

The steel used in the present study is designated at INRAFM steel and the chemical composition of the steel is given in Table 1. The steel was subjected to austenisation at 1253 K for 30 min and subjected to air cooling (normalizing). The normalized steel was also tempered at 1033 K for 1 h.

Table 1. Chemical composition of INRAFM steel (wt%).

Element	Concentration	Element	Concentration
Cr	9.03	**Ta**	0.06
C	0.126	**N**	0.03
Mn	0.56	**O**	0.002
V	0.24	**P**	<0.002
W	1.38	**S**	<0.001

Preliminary microstructural analysis was carried out in scanning electron microscope (Philips XL30, F.E.I. Company: Hillsboro, OR, USA). Angle dispersive X-ray diffraction (ADXRD) experiments were carried out using beam line (BL-12) at an Indian synchrotron source (Indus 2), at Raja Ramanna Centre for Advanced Technology (RRCAT), Indore, since the high brilliance of synchrotron source compared to conventional source would enable the identification of phases of volume fraction less than 1%. A monochromatic X-ray beam ($\lambda/\Delta\lambda$ = 8000 at ~8 keV) of photon energy 13.081 KeV (λ = 0.9478 A°) or 16 keV and beam size 300 × 300 μm^2 was used. NIST LaB_6 powder was used for calibration of wavelength in transmission mode. Synchrotron XRD experiments were carried out on thin foils of the steels (~100 μm thick) in the transmission mode using MAR Research image plate (MAR 345 DTB). Image plate data was converted to I vs. 2 θ plot using FIT2D software. Mossbauer spectroscopy was carried out on thin foils (~100 μm thick) of the steels, with Co^{57} source dispersed in Rh matrix as the probe with a Wiesel spectrometer in transmission geometry. Thin foil specimens for TEM were prepared by conventional mechanical thinning followed by electrolytic polishing with a Struers jet thinning apparatus (Tenupol-5) with appropriate electrolyte, while carbon extraction replica specimens were prepared by picking up the thin film of carbon deposited on the steel surface immersed in Villela's reagent followed by rinsing in methanol transmission electron microscopy observations were carried out with Philips CM200 analytical transmission electron microscopy (ATEM) at an acceleration voltage of 200 kV. EDX analysis was carried out with the Oxford EDS system attached to the ATEM. Orientation imaging in TEM was carried out with an automated crystal orientation mapping (ACOM) system from M/s. Nanomegas SPRL, Brussels, Belgium.

3. Results

Secondary electron (SE) and back scattered electron (BSE) images of the normalized steel are shown in Figure 1a,b. It is observed that the structure is martensitic in the normalized steel. Presence of undissolved primary carbides is also observed in the normalized steel from Figure 1b. Detailed selected area diffraction (SAD) and energy dispersive X-ray (EDX) analysis of the precipitates seen in TEM micrograph in Figure 1c confirmed that the precipitates are Cr, W rich $M_{23}C_6$ and Ta, V rich MX as shown in Figure 1d. In addition, the presence of fine, acicular carbides has been noted in a few wide laths of the normalized steel (Figure 2a), which were identified to be Fe rich M_3C carbides as shown in Figure 2b,c.

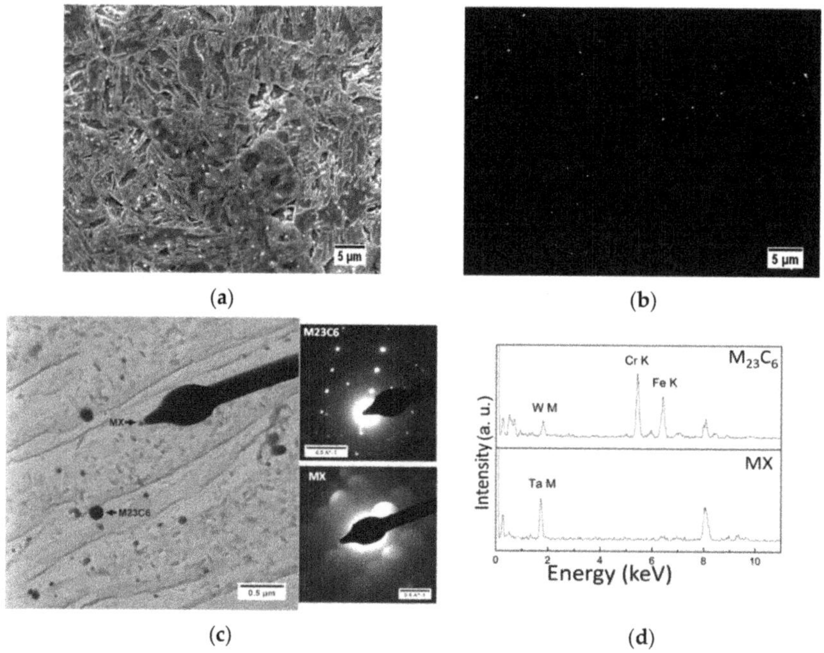

Figure 1. Microstructure of 1.4W-0.06Ta steel normalized at 1253 K: (**a**) SE image showing martensite; (**b**) BSE image showing undissolved carbides; (**c**) carbon extraction replica BF image along with diffraction patterns from $M_{23}C_6$ and MX along $[5\,4\,\bar{5}]$ and $[\bar{1}\,1\,2]$ zone axis respectively; (**d**) EDX spectra from undissolved $M_{23}C_6$ and MX carbides.

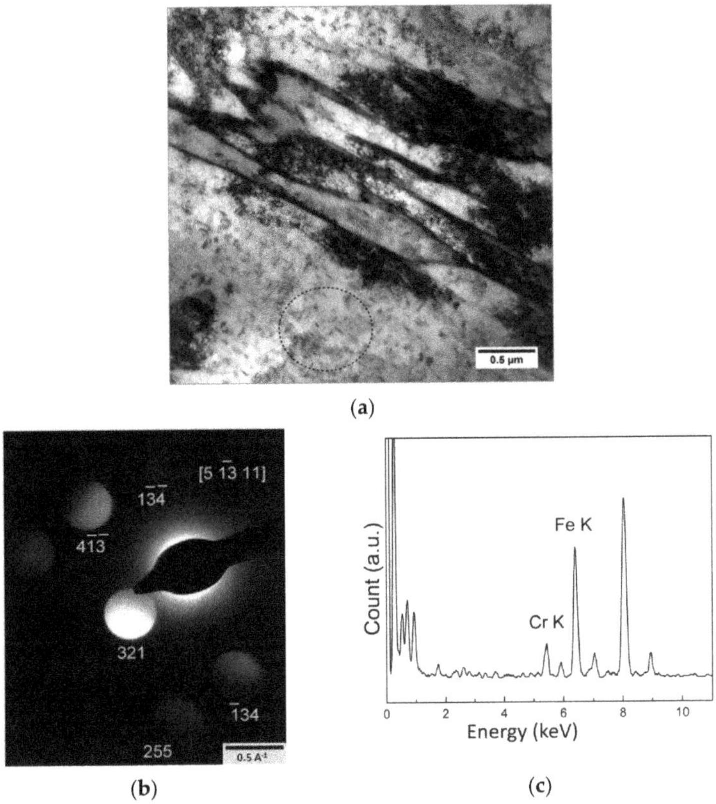

Figure 2. (**a**) Bright field image of normalised1.4W-0.06Ta steel showing fine carbides (circled) in a wide lath; (**b**) micro diffraction pattern from a precipitate along -$[5\ \overline{13}\ 11]$ zone axis of M_3C; (**c**) typical EDS spectrum of these carbides from carbon extraction replica of the same sample, showing Fe enrichment.

Microstructural features of the normalized and tempered steel are shown in Figure 3a–d. It is observed that the microstructure is tempered martensite with the presence of higher amounts of Cr rich $M_{23}C_6$ at the boundaries of prior austenite grains and martensitic laths and intralath MX precipitates enriched with Ta, V, which is in agreement with results available in literature on these steels [14]. However, contrary to the presence of Fe rich M_3C in the normalized steel, no signature for M_3C could be found in the tempered steel. This suggests the metastable nature of M_3C.

Conventional laboratory XRD spectra of the normalized as well as normalized and tempered steels showed the presence only α-bcc peaks, corresponding to the martensite/ferrite matrix, due to the low amounts of precipitates in the tempered steel. Methods such as electrolytic phase extraction can enable the identification of low amounts secondary phases with the matrix interference even by laboratory X-ray sources [20] or alternatively, synchrotron XRD is well known to be suitable for phase identification with low detection limits [21]. ADXRD spectra of the normalized as well as the tempered steels are presented in Figure 4a. Since the carbon concentration of the steel used in the present study is low, it is assumed that the lattice parameter of martensite is not significantly different from BCC α-Fe. Apart from the major phase α-Fe, peaks corresponding to the secondary phases $M_{23}C_6$ and MX precipitates could also be indexed. However, the intensity ratio of the peaks did not follow the JCPDS data, due to certain experimental difficulties such as static nature of the sample, intensity saturation of the image plate, etc. In the ADXRD spectrum of the normalized steel, (Figure 4b), the (110)α peak of highest intensity is asymmetrical and could be resolved into two peaks, the one with a lower intensity identified as (111) peak of austenite. The other peaks of austenite could not be unambiguously indexed, since they were overlapping with that of the undissolved $M_{23}C_6$ and MX precipitates. On the other hand, in the normalized and tempered steel, such a deconvolution of $(110)_\alpha$ peak was not possible, which clearly indicates the absence of austenite after tempering.

Figure 3. Microstructure of normalised and tempered 1.4W-0.06Ta steel showing tempered martensite: (**a**) SE image showing extensive precipitation of carbides along PAGB and lath boundaries; (**b**,**c**) TEM micrographs showing interlath $M_{23}C_6$, intra lath MX precipitates; (**d**) EDX spectra from Cr rich $M_{23}C_6$ carbides and Ta, V rich MX precipitates.

Figure 5a,b shows the Mossbauer spectra of normalized as well as normalized and tempered steels. It is observed that the steel in both the conditions exhibit a six-line pattern, which could be further resolved into three six-line patterns. Such resolved components are due to a complex interplay of the alloying elements, which is beyond the scope of this study. The presence of a six-line pattern is mainly attributed to the predominant ferro magnetic bcc ferrite. Interestingly, in the normalized steel, in addition to the six-line patterns, spectrum could be deconvoluted to show the presence of a single line pattern of low intensity (arrow marked in Figure 5a). Such a singlet clearly suggests the presence of a non-magnetic cubic phase, which could be retained austenite or the undissolved $M_{23}C_6$ and MX precipitates, since all these phases are known to possess a fcc structure and also paramagnetic. However, no such singlet was observed in the normalized and tempered steel, where the amount of $M_{23}C_6$ and MX precipitates, is higher than in the normalized steel. Hence, the singlet observed in the normalized steel should correspond to retained austenite of a low volume fraction [9,22] rather than due to the presence of undissolved carbides.

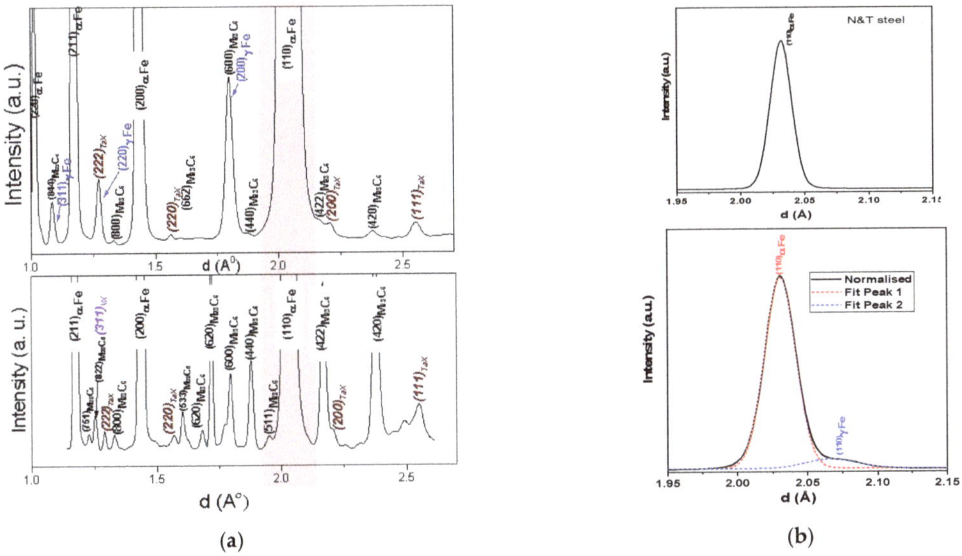

Figure 4. (**a**) ADXRD spectra of Normalised and normalised and tempered steel showing different phases; (**b**) (110) peak of ferrite (BCC) and (111) peak of austenite (FCC) phases resolved in normalised steel (top) after deconvolution; (110) peak of ferrite (BCC) in normalised and tempered steel.

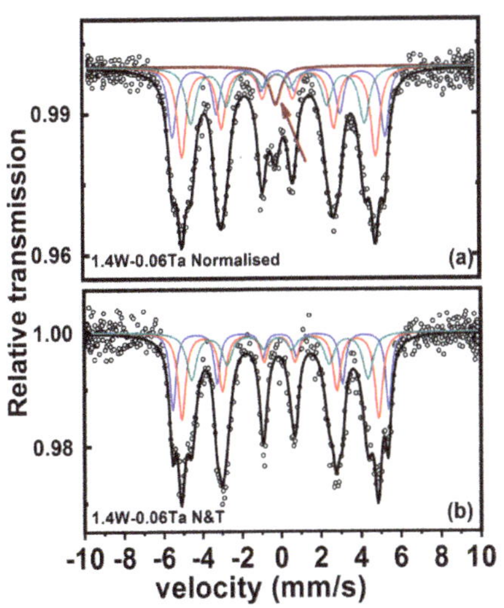

Figure 5. Mossbauer spectra of the 1.4W-0.06Ta steel in: (**a**) normalized condition showing the presence of a single line pattern (arrow marked) due to low amount of retained austenite in addition to six-line pattern due to ferromagnetic ordering; (**b**) normalized and tempered condition (bottom) showing only ferromagnetic ordering.

Conventional TEM observations could not unambiguously confirm the presence of retained austenite in the normalized steel. Hence a newly emerging technique automated crystal orientation mapping in TEM (ACOM-TEM), was employed. Crystallographic data for α ferrite, γ austenite and $M_{23}C_6$ were given as input for analysis. Figure 6a,b shows the resultant index and reliability maps of the normalized steel, while Figure 6c shows the orientation map from the same region. Multiple orientation of martensite is quite evident from Figure 6c, which can be attributed to a fast-cooling rate adopted during normalizing. The phase map presented in Figure 6d showed the presence of ~2% retained austenite.

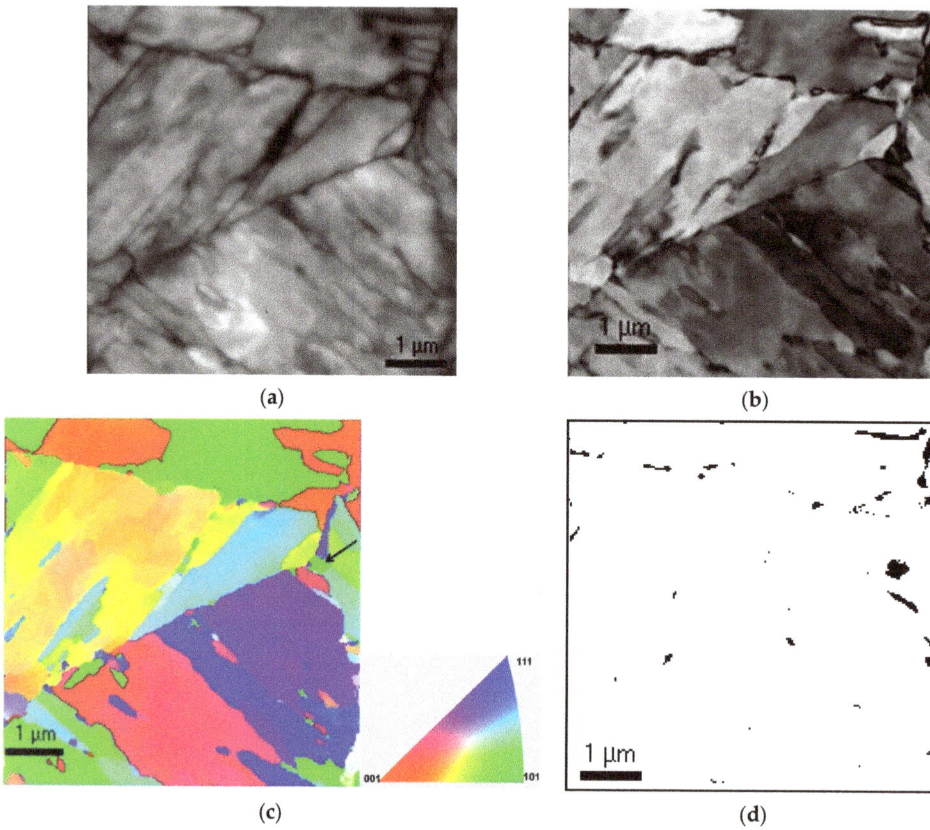

Figure 6. Orientation imaging of normalized 1.4W-0.06Ta steel showing: (**a**) index map; (**b**) reliability map; (**c**) orientation map showing random orientation of martensite laths; (**d**) phase map showing retained γ.

4. Discussion

Formation of martensite in 9Cr ferritic/martensitic steels air cooled at the rate of ~2 K/min is well reported [23]. Transformation of Fe from FCC to BCC is known to follow the Bain, Kurdjumov–Sachs (K-S), Nishiyama–Wassermann (N-W) or Greninger–Troiano (G-T) orientation relationship (OR) [24]. K-S and N-W OR are more frequently reported in low C steels. However, the more common factors, which can influence on OR are thermo-mechanical processing and chemical composition [24,25]. In low carbon ferritic/martensitic steels, $\gamma \rightarrow \alpha'$ transformation reportedly follows K-S OR [26,27],which can result in 24 crystallographic variants of α from a given γ orientation. Hence in this study also, it is expected that martensite forms obeying the K-S OR.

From Figure 6c, it is clearly observed that several orientations of the packets of martensite nucleated from the 4 {111} planes of γ as indicated by the legend in Figure 6c, the standard stereographic triangle of a cubic system. The observation of many orientations within each austenite grain suggests that there is no variant selection, which is due to a relatively fast cooling rate adopted.

Retained austenite in modified 9Cr-1Mo weldments is reported to obey both K-S and N-W orientation relationship [15]. However, Duanjun Sun et al. [28] observe that even when the amount of retained austenite is less, there can be a deviation in the orientation of austenite above M_s, which may be inevitable as a result of strain associated with martensitic transformation. It is reported that thin films of retained austenite between laths of low alloy martensite are mechanically stabilized at the austenite-martensite interface by transformation induced accommodation defects [15]. However, a high degree of mutual accommodation of transformation strains between adjacent variants of martensite with a K-S orientation relationship such as in twin related martensite can suppress the retention of inter-lath austenite [29]. Hence, the observation of retained austenite along with martensite in the normalized steel without any variant selection of martensite in this study could be attributed to its mechanical stabilization associated with the transformation strain.

It is known that γ to α' transformation is significantly influenced by factors such as (a) cooling rate, (b) alloying additions, and (c) homogeneity of austenite. An earlier study of differential scanning calorimetry analysis of this steel austenitized and cooled at different rates clearly showed that at cooling rates less than 40 K/min, there is a cooling rate dependence of the M_s and M_f temperatures [30], with M_s and M_f temperatures decreasing with increase in cooling rate. In addition, it is known both M_s and M_f temperatures decrease with the addition of ferrite stabilizers W and Ta, which are also strong carbide formers [6]. The dissolution temperature of $M_{23}C_6$ and MX precipitates in this steel was determined to be higher than 1342 and 1570 K respectively [30]. Hence, in this steel with 1.4% W, it can be inferred that normalizing at 1253 K is insufficient for complete dissolution of these carbides as shown in Figure 1. Such incomplete dissolution of the primary carbides can lead to the inhomogeneity of the austenite, which would increase with W or Ta content of the steel [12]. Further detailed analysis of martensitic lath structure in TEM showed two different types of lath structure [12]. Such a precipitation of Fe rich M_3C was shown to occur in a few wide laths, while the laths with narrow width showed no precipitation. This had been understood to be due to the inhomogeneous austenite around the undissolved carbides—one lean w.r.t carbon and the other enriched with carbon, since it is known that carbon migration is possible even at lower temperatures also [31]. Carbon lean austenite has a relatively higher M_s and M_f temperatures leading to the partial transformation of γ to $\alpha + carbides$ (similar to bainite) above M_s, [12,32] as shown in Figure 2. Variation in lath size and the carbide precipitation indicates that the stability of the inhomogeneous austenite is influenced by the cooling rate. The observation of the austenite in the normalized steel has been unambiguously confirmed by various experimental techniques, which could be due to an incomplete martensitic transformation of the carbon enriched austenite. Presence of retained austenite as thin films at martensitic lath boundaries in the weldments of this class of steels has been reported in literature [33]. Hence the transformation sequence of the steel during normalizing can be summarized as

$$Austenite + undissolved\ carbides \rightarrow martensite\ (\alpha') + \alpha + carbides + traces\ of\ retained\ austenite\ (\gamma_r)$$

Absence of Fe rich carbides in the tempered steel proves that the M_3C present after normalization is unstable. It is well reported that the stable carbides in these steels after tempering in the temperature range of 750–780 °C are $M_{23}C_6$ and MX [18,34]. In addition, no signature of austenite could be found in the tempered steel. Hence, it can be inferred that the observed microstructure containing *martensite* + α + *carbides* + γ_r is due to the complex interplay of micro chemical variation/inhomogeneity in the alloy even after normalizing at 1253 K for 30 min. This is also attributed to the sluggish diffusion of heavy elements such as W and Ta [35] in the steel. Further to support the fact of sluggish kinetics, it is reported in an earlier work that the number density of carbides in the steel increases with increase in W content [36].

5. Conclusions

Microstructural investigation on 9Cr-1.4W-0.12C Ferritic martensitic steel was performed in the present work. The salient results of the study are as follows:

- Conventional TEM analysis of normalized steel shows a martensitic structure with few wide laths containing Fe rich M_3C carbides;
- Detailed analysis of ADXRD, orientation imaging in TEM and Mossbauer studies proved the presence of retained austenite in normalised steel;
- The presence of M_3C and retained austenite in the normalized steel with a martensitic microstructure is understood to be due to the inhomogeneity of austenite with incomplete dissolution of primary carbides during normalizing treatment;
- Absence of retained austenite and M_3C in the tempered steel suggested their metastable nature, which could have transformed to stable $M_{23}C_6$ and MX precipitates on tempering.

Author Contributions: Conceptualization, methodology, R.M., S.S. and M.V.; data curation, formal analysis, investigation, R.M., R.K., L.H.S., R.G., M.N.S. and A.K.S.; writing—original draft preparation, R.M., R.K. and S.S.; writing—review and editing, R.M., R.K., S.S. and R.G.; project administration, S.S., M.V., S.K.D. All authors have read and agreed to the published version of the manuscript.

Funding: This research received no external funding.

Institutional Review Board Statement: Not applicable.

Informed Consent Statement: Not applicable.

Data Availability Statement: The data that support the findings of this study are not available at present, as it is a part of the ongoing research.

Acknowledgments: The authors wish to thank Arup Dasgupta, Physical Metallurgy Division, S. Raju, Metallurgy and Materials Group and B. Venkatraman, Indira Gandhi Centre for Atomic Research for their encouragement and support during this work.

Conflicts of Interest: The authors declare no conflict of interest.

References

1. Klueh, R.L. Elevated temperature ferritic and martensitic steels and their application to future nuclear reactors. *Int. Mater. Rev.* **2005**, *50*, 287–310. [CrossRef]
2. Harries, D.R. *High-Chromium Ferritic and Martensitic Steels for Nuclear Applications*; ASTM: West Conshohocken, PA, USA, 2001.
3. Abe, F. Precipitate design for creep strengthening of 9% Cr tempered martensitic steel for ultra-supercritical power plants. *Sci. Technol. Adv. Mater.* **2008**, *9*. [CrossRef]
4. Asada, Y.; Dozaki, K.; Ueta, M.; Ichimiya, M.; Mori, K.; Taguchi, K.; Kitagawa, M.; Nishida, T.; Sakon, T.; Sukekawa, M. Exploratory research on creep and fatigue properties of 9Cr-steels for the steam generator of an FBR. *Nucl. Eng. Des.* **1993**, *139*, 269–275. [CrossRef]
5. Zhu, L.; Liu, X.; Fan, P.; Liu, J. A Study of Microstructure Evolution During Creep of 9Cr-1Mo Steel Using Ultrasonic and Hardness Measurements. *J. Mater. Eng. Perform.* **2019**, *28*, 2348–2355. [CrossRef]
6. Li, X.; Yan, Q.; Ma, R.; Wang, H.; Ge, C. First results of characterization of 9Cr-3WVTiTaN low activation ferritic/martensitic steel. *J. Iron Steel Res. Int.* **2010**, *17*, 57–62. [CrossRef]
7. Gao, Y.; Wang, Z.; Liu, Y.; Li, W.; Liu, C.; Li, H. Diffusion Bonding of 9Cr Martensitic/Ferritic Heat-Resistant Steels with an Electrodeposited Ni Interlayer. *Metals* **2018**, *8*, 1012. [CrossRef]
8. Kim, T.K.; Kim, S.H.; Lee, B. Effects of an intermediate heat treatment during a cold rolling on the tensile strength of a 9Cr-2W steel. *Ann. Nucl. Energy* **2009**, *36*, 1103–1107. [CrossRef]
9. Besoky, J.I.; Danon, C.A.; Ramos, C.P. Retained austenite phase detected by Mössbauer spectroscopy in ASTM A335 P91 steel submitted to continuous cooling cycles. *J. Mater. Res. Technol.* **2019**, *8*, 1888–1896. [CrossRef]
10. Wu, Q.S.; Zheng, S.H.; Huang, Q.Y.; Liu, S.J.; Han, Y.Y. Continuous cooling transformation behaviors of CLAM steel. *J. Nucl. Mater.* **2013**, *442*, S67–S70. [CrossRef]
11. Ni, M.; Wang, J.; Liu, J.; Wang, X.; Zhang, K.; Du, C. Microstructure and Mechanical Properties of P91 Steel during Heat Treatment: The Effect of the Cooling Speed during the Normalization Stage. *J. Mater. Eng. Perform.* **2012**, *30*, 2329–2340. [CrossRef]
12. Ravikirana Mythili, R.; Raju, S.; Saroja, S.; Jayakumar, T.; Rajendra Kumar, E. Decomposition modes of austenite in 9Cr-W-V-Ta reduced activation ferritic-martensitic steels. *Mater. Sci. Technol.* **2015**, *31*, 448–459. [CrossRef]

13. Tan, L.; Hoelzer, D.T.; Busby, J.T.; Sokolov, M.A.; Klueh, R.L. Microstructure control for high strength 9Cr ferritic—Martensitic steels. *J. Nucl. Mater.* **2012**, *422*, 45–50. [CrossRef]
14. Haarmann, K.; Vaillant, J.; Bendick, W.; Arbab, A. *The T91/P91 Book*; Vallourec Mannesmann Tubes: Duesseldorf, Germany, 2002; p. 69.
15. Shiue, R.K.; Lan, K.C.; Chen, C. Toughness and Austenite Stability Of Modified 9Cr-1Mo welds after tempering. *Mater. Sci. Eng. A* **2000**, *287*, 10–16. [CrossRef]
16. Santella, M.L.; Babu, S.S.; Swindeman, R.W.; Specht, E.D. In-situ characterization of austenite to martensite decomposition in 9Cr-1Mo-V steel welds. In Proceedings of the Materials Science and Technology 2003 Meeting, Chicago, IL, USA, 9–12 November 2003; pp. 247–256.
17. Ning, B.; Zhou, X.; Shi, Q.; Liu, Y.; Zhao, J.; Zhang, Z. Relationship between austenite stability and martensite formation in modified 9Cr-1Mo steel. *Int. J. Mater. Res.* **2014**, *105*, 232–239. [CrossRef]
18. Fedorova, I.; Kostka, A.; Tkachev, E.; Belyakov, A.; Kaibyshev, R. Tempering behavior of a low nitrogen boron-added 9%Cr steel. *Mater. Sci. Eng. A* **2016**, *662*, 443–455. [CrossRef]
19. Zhang, Y.; Gu, J.; Han, L.; Shen, G.; Li, C. Thermal decomposition characteristics of retained austenite and its influence on impact toughness of B-containing 9Cr1Mo1Co(FB2) steel during the two-step tempering. *J. Mater. Res. Technol.* **2021**, *12*, 2462–2475. [CrossRef]
20. Gao, Q.; Zhang, Y.; Zhang, H.; Li, H.; Qu, F.; Han, J.; Lu, C.; Wu, B.; Lu, Y.; Ma, Y. Precipitates and Particles Coarsening of 9Cr-1.7W-0.4Mo-Co Ferritic Heat-Resistant Steel after Isothermal Aging. *Sci. Rep.* **2017**, *7*, 5859. [CrossRef] [PubMed]
21. Suwanpinij, P. The Synchrotron Radiation for Steel Research. *Adv. Mater. Sci. Eng.* **2016**, *2016*, 2479345. [CrossRef]
22. Carrizo, D.A.; Besoky, J.I.; Luppo, M.; Danon, C.; Ramos, C.P. Characterization of an ASTM A335 P91 ferritic-martensitic steel after continuous cooling cycles at moderate rates. *J. Mater. Res. Technol.* **2019**, *8*, 923–934. [CrossRef]
23. Saroja, S.; Vijayalakshmi, M.; Raghunathan, V.S. Influence of solution treatment on the microstructure of a 9wt.%Cr-lwt.%Mo-0.07wt.%C steel. *Mater. Sci. Eng. A* **1992**, *154*, 59–67. [CrossRef]
24. Suikkanen, P.P.; Cayron, C.; DeArdo, A.J.; Karjalainen, L.P. Crystallographic analysis of martensite in 0.2C-2.0Mn-1.5Si-0.6Cr steel using EBSD. *J. Mater. Sci. Technol.* **2011**, *27*, 920–930. [CrossRef]
25. Kwon, E.P.; Fujieda, S.; Shinoda, K.; Suzuki, S. Martensitic transformation and texture in novel bcc Fe-Mn-Al-Ni-Cr alloys. *Procedia Eng.* **2011**, *10*, 2214–2219. [CrossRef]
26. Dash, M.K.; Karthikeyan, T.; Saroja, S.; Vijayalakshmi, M. Restitution of prior-austenite grain orientation by microtexture analysis of tempered martensite structure in 9Cr-1Mo ferritic steel. *Mater. Sci. Forum* **2012**, *702–703*, 880–883. [CrossRef]
27. Kitahara, H.; Ueji, R.; Tsuji, N.; Minamino, Y. Crystallographic features of lath martensite in low-carbon steel. *Acta Mater.* **2006**, *54*, 1279–1288. [CrossRef]
28. Sun, D.; Li, C.; Xue, X.; Liu, Y.; Guo, Z.; Gu, J. Optimization scheme of the orientation relationship from crystallographic statistics of variants and its application to lath martensite. *Mater. Des.* **2020**, *195*, 109022. [CrossRef]
29. Bhadeshia, H.K.D.H. Theory and Significance of Retained Austenite in Steels. Ph.D. Thesis, University of Cambridge, Cambridge, UK, March 1980.
30. Kirana, R.; Raju, S.; Mythili, R.; Saibaba, S.; Jayakumar, T.; Rajendra Kumar, E. High-Temperature Phase Stability of 9Cr-W-Ta-V-C Based Reduced Activation Ferritic-Martensitic (RAFM) Steels: Effect of W and Ta Additions. *Steel Res. Int.* **2015**, *86*, 825–840. [CrossRef]
31. Speer, J.; Matlock, D.K.; De Cooman, B.C.; Schroth, J.G. Carbon partitioning into austenite after martensite transformation. *Acta Mater.* **2003**, *51*, 2611–2622. [CrossRef]
32. Liu, C.; Zhang, D.; Liu, Y.; Wang, Q.; Yan, Z. Investigation on the precipitation behavior of M3C phase in T91 ferritic steels. *Nucl. Eng. Des.* **2011**, *241*, 2411–2415. [CrossRef]
33. Shirane, T.; Tsukamoto, S.; Tsuzaki, K.; Adachi, Y.; Hanamura, T.; Shimizu, M.; Abe, F. Ferrite to austenite reverse transformation process in B containing 9%Cr heat resistant steel HAZ. *Sci. Technol. Weld. Join.* **2009**, *14*, 698–707. [CrossRef]
34. Kipelova, A.Y.; Belyakov, A.N.; Skorobogatykh, V.N.; Shchenkova, I.A.; Kaibyshev, R.O. Tempering induced structual changes in steel 10Kh9K3V1M1FBR and their effect on the mechanical properties. *Met. Sci. Heat Treat.* **2010**, *52*, 100–110. [CrossRef]
35. Kim, H.K.; Lee, J.W.; Moon, J.; Lee, C.H.; Hong, H.U. Effects of Ti and Ta addition on microstructure stability and tensile properties of reduced activation ferritic/martensitic steel for nuclear fusion reactors. *J. Nucl. Mater.* **2018**, *500*, 327–336. [CrossRef]
36. Mythili, R.; Saroja, S. Influence of Tungsten and Tantalum Content on Evolution of Secondary Phases in 9Cr RAFM Steels: An Experimental and Computational Study. *Metall. Mater. Trans. A* **2017**, *48*, 3880–3891. [CrossRef]

symmetry

MDPI

Article

Characterization of Microstructure of Crept Nb and Ta-Rich γ-TiAl Alloys by Automated Crystal Orientation Mapping and Electron Back Scatter Diffraction

Vajinder Singh [1], Chandan Mondal [2], Rajdeep Sarkar [1], Satabdi Roy [1], Chiptalluri Mohan Omprakash [3] and Partha Ghosal [1,*]

1 Electron Microscopy Group, Defence Metallurgical Research Laboratory, Kanchanbagh, Hyderabad 500058, India; vajindersingh.dmrl@gov.in (V.S.); rajdeepsarkar.dmrl@gov.in (R.S.); satabdi.dmrl@gov.in (S.R.)

2 Rolling & Formability Group, Defence Metallurgical Research Laboratory, Kanchanbagh, Hyderabad 500058, India; c-mondal.dmrl@gov.in

3 Mechanical Behavior Group, Defence Metallurgical Research Laboratory, Kanchanbagh, Hyderabad 500058, India; cmo.dmrl@gov.in

* Correspondence: parthaghosal.dmrl@gov.in

Abstract: Understanding of the creep behavior Nb and Ta-rich γ-TiAl alloys plays a crucial role towards realization of their potential applications. The present article reports the evolution of microstructural features in the crept γ-TiAl-based Ti-5Al-8Nb-2Cr-0.2B and Ti-45Al-8Ta-0.2C-0.2B-0.2C alloys. Structural characterizations have been carried out using automated crystal orientation mapping (ACOM) along with precession electron diffraction (PED) in a transmission electron microscope, in conjunction with electron back-scattered diffraction (EBSD) in a scanning electron microscope (SEM) and transmission electron microscopy (TEM). Creep behavior of the fourth generation γ-TiAl-based alloys has been comparatively investigated under constant load tensile creep tests performed in the temperature range from 800–850 °C and applied stresses range of 125–200 MPa. It has been demonstrated that the ACOM with PED technique has accurate and reliable diffraction pattern recognition and higher spatial resolution, and supplements effectively the conventional EBSD technique for characterization of complex microstructural features evolved during creep of multiphase ($\gamma + \alpha_2 + \beta$)-based TiAl alloys. The results show that the Nb and Ta additions have distinctly different effects on the microstructural instability and phase transformation during the creep deformation. The formation of the Ta-rich intermetallic phase (Ti_4Al_3Ta, the so-called τ phase) has been detected preferentially along the colony and the γ-α_2 interphase boundaries in the Ta-rich alloy, whilst its isomorph, Ti_4Al_3Nb intermetallic, has hardly been detected in the Nb-rich alloy. Implications of τ-phase formation and related microstructural instabilities have been discussed with respect to the creep behavior of the two alloys.

Keywords: γ-TiAl alloys; creep; microstructure; ACOM-PED; EBSD

1. Introduction

γ-TiAl based alloys have the potential to replace the existing high-temperature materials used in the structural application of gas turbine and automotive industries. The alloying addition to the γ-TiAl alloys significantly modifies the microstructure in terms of single-phase, two-phase, or multiphase combinations, which further control the high-temperature mechanical properties such as creep. Typically, solidification of γ-TiAl alloys in the composition range of Ti-(40–50 at %) Al results in a near or fully lamellar microstructure which can be further modified to duplex and equiaxed types by heat treatment and thermo-mechanical processing [1]. Systematic research on alloy design has shown that the lamellar microstructure exhibits better creep resistance as compared to other morphologies. However, the structural instability in terms of the microstructural coarsening and spheroidization do occur in fully lamellar structure especially when exposed to high temperature for a long time. Microstructure coarsening is caused via the diffusion mechanism [2,3]. In addition, long heat treatment also activates precipitation and phase transformation which either improves or deteriorates the creep resistance. To improve the creep resistance and structural stability of γ-TiAl based alloys for aerospace and automotive applications, the alloys are required to be modified by the addition of alloying elements such as Nb, Ta in the range of (5–10 at %) and Cr, Mo, and V (2–4 at %) and B, C (0.1–0.2 at %) which reduce the diffusion rate, provide solid solution hardening, and refine the microstructure [1,4]. In recent years, investigations have been reoriented toward Nb-based and Ta-based γ-TiAl alloys, the so-called third-generation, and fourth-generation alloys. These alloys show improvement in mechanical strength and creep resistance. Furthermore, Nb and Ta additions to TiAl alloys increase the oxidation resistance significantly and reduce the stacking fault energy [5,6]. In recent studies, Ti—45Al-8Nb single crystal (poly-synthetically twinned) and Ti-46Al-8Ta (at %) air-hardenable alloy show superior creep resistance properties, but poor hot workability limits their application potentials [7,8]. This limitation can be overcome by introducing a novel alloy design concept called the "β-solidification route" [9,10]. The β-solidifying alloys contain either high Nb or Ta element along with β-stabilizers (such as Mo, W, Cr, Mn, etc.), and B and/or C as refiners. However, the retained β(B2) phase also has a deleterious effect on creep resistance when present in high volume fractions in γ-TiAl based alloys [11]. From the foregoing discussion it is amply clear that understanding the role of microstructural constituents and their stability play a crucial role in controlling the creep behavior. The present study aims to characterize the microstructural evolution and stability during the creep deformation of the β(B2)-stabilized Ti-45Al-8Nb-2Cr-0.2B and Ti-45Al-8Ta-2Cr-0.2B-0.2C alloys.

2. Materials and Methods

Gamma TiAl alloys having nominal compositions of Ti-45Al-8Nb-2Cr-0.2B and Ti-45Al-8Ta-2Cr-0.2B-0.2C (at %) were fabricated in vacuum arc furnace in the form of pancake. The carbon was avoided in high-Nb alloy, as carbon addition leads to the formation of Ti_3AlC or Ti_2AlC phases in these alloys, which causes brittleness in the alloy system and deteriorates the workability, as reported by Gabrisch et al. [12]. Contrarily, the formation of the carbide phases in high-Ta alloy has still not been verified. Creep behavior was evaluated on the as-cast condition in the temperature range of 800–850 °C at stresses ranging from 150 to 200 MPa. Compression creep tests were carried out under constant load conditions in air using STAR creep test Systems having a lever ratio of 1:20 using parallelepiped samples of 10 (L) × 5 (w) × 5 (t) mm.

The microstructural evaluation of the cast alloy was carried out in pre- and post-creep tested specimens by Field Emission Scanning Electron Microscope (FESEM, model: Supra 55 and Cross Beam 350), Energy Dispersive X-ray spectroscopy (EDS), and Electron Backscatter Diffraction (EBSD) Oxford make in FEG SEM Carl Zeiss make. XRD analysis was done by using a diffractometer equipped with an X-ray tube with a rotating Cu anode at 12 kW. EBSD samples were prepared using standard metallographic methods and final polishing was done in a Buehler vibratory polisher for 12 h. The acquisition of orientation data has been done with an Oxford Nordlys detector and analyses have been carried out using TSL 8.0 software. The raw data have been further processed by removing the pseudo-symmetry effect [13]. Moreover, an advanced technique called Automated Crystal Orientation Mapping (ACOM) along with precession electron diffraction (PED) in TEM (NanoMegas ASTAR system, Brussel, Belgium) was used [14,15]. A precession angle of 0.5° with 100 Hz frequency was employed for ACOM maps to characterize finer microstructural features involving nano level lath and precipitates in the crept samples. A step size of 20 nm, and camera exposure time of 20 ms were used for all the maps. ACOM, along with PED study and conventional imaging, were undertaken using the TEM (Tecnai G2 Model, FEI make, Hillsboro, OR, USA), and the spatial resolution was kept at 20 nm for the present experiment. The raw data was converted through the TSL software for further analysis. ACOM analysis is similar to EBSD; however, in the former method the grain orientation and the phases are identified through the diffraction spots/patterns with very high spatial resolution in sub-nanometer range. The samples for TEM and PED were prepared by Precision Ion Polishing System (PIPS, make: Gatan, Pleasanton, CA, USA) using 4 kV and ±4° gun tilt (upper and lower) followed by final polishing with 3 kV and ±3° gun tilt.

3. Results and Discussion

3.1. Microstructure of the As-Cast Materials

The microstructures of the as-cast Ti-45Al-8Nb-2Cr-0.2B and Ti-45Al-8Ta-2Cr-0.2B-0.2C alloys are shown in Figure 1a–e, respectively. The back-scattered SEM micrographs show the presence of a bright region near the colony boundaries. These bright regions are due to retained β(B2) phase formed by segregation of Nb, Ta, and Cr during solidification, which was also confirmed through EDS microanalyses (not shown here).

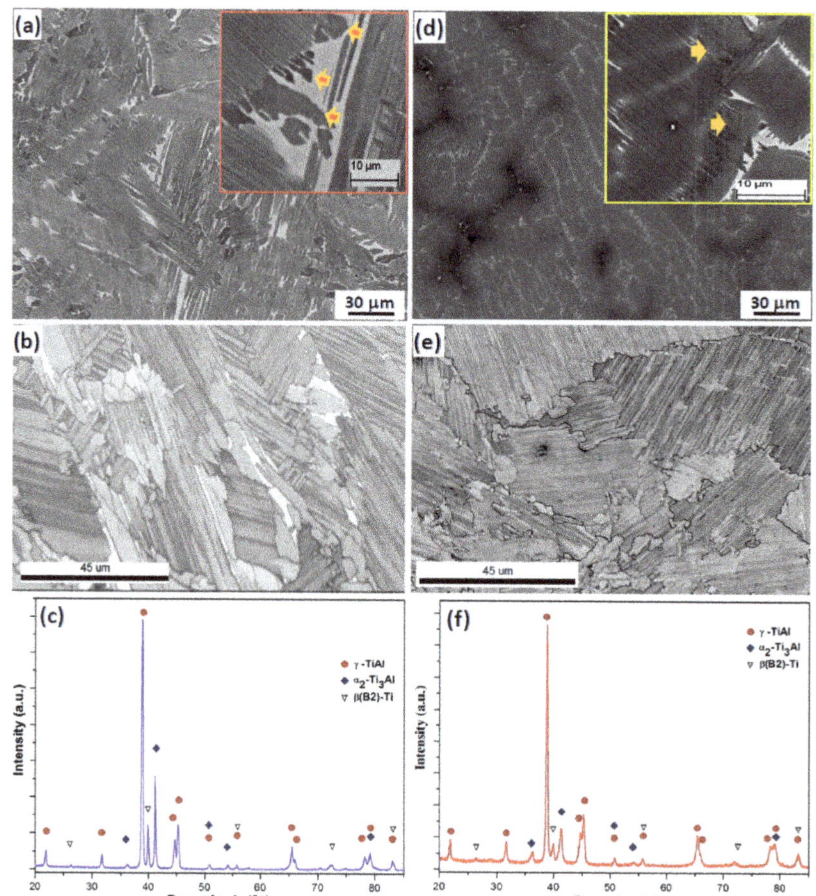

Figure 1. Backscattered SEM image, EBSD-band contrast image, and corresponding XRD pattern of the as-cast pancakes of (**a–c**) Ti-45Al-8Nb-2Cr-0.2B alloy, and (**d–f**) Ti-45Al-8Ta-2Cr-0.2B-0.2C alloy.

At low magnification, the microstructure shows a non-uniform distribution of the colony size and some intermittent γ-TiAl grains. EBSD band contrast images (Figure 1b,e) of both the alloys further illustrates the as-cast microstructure. The colony size for both the alloys was found to vary between 40–100 μm. Figure 1b shows that the microstructure is predominantly lamellar, and that the Ta-rich alloy has a much finer lamellar spacing compared to the Nb-rich alloy (Figure 1e). For both the alloys, the retained β-phase is mostly observed in the colony boundaries, and frequently punctuated by transformed γ-TiAl phase. One of the prominent differences between the as-cast microstructures of Nb-rich and Ta-rich alloys is the presence of strong inter-dendritic negative-segregation of Ta, whereas in the case of Nb-rich alloy, such segregation is generally absent. X-ray diffraction (XRD) study has been further carried out to identify the phases present. The representative XRD patterns of the alloys are shown in Figure 1c,f. The phase analysis of the major peaks reveals the presence of three constituent phases, namely α_2, γ, and β. These results are consistent with the findings of the SEM analysis.

3.2. Compressive Creep Behavior

Representative creep curves for the Nb- and Ta-rich alloys in the as-cast condition at 800–850 °C and an applied stress level of 125–200 MPa are shown in Figure 2. At lower temperature (800 °C), two distinct stages of creep can be noted in the creep curve i.e., a primary creep stage and a well-defined secondary or steady-state creep region. The primary stage becomes increasingly shorter at higher temperatures.

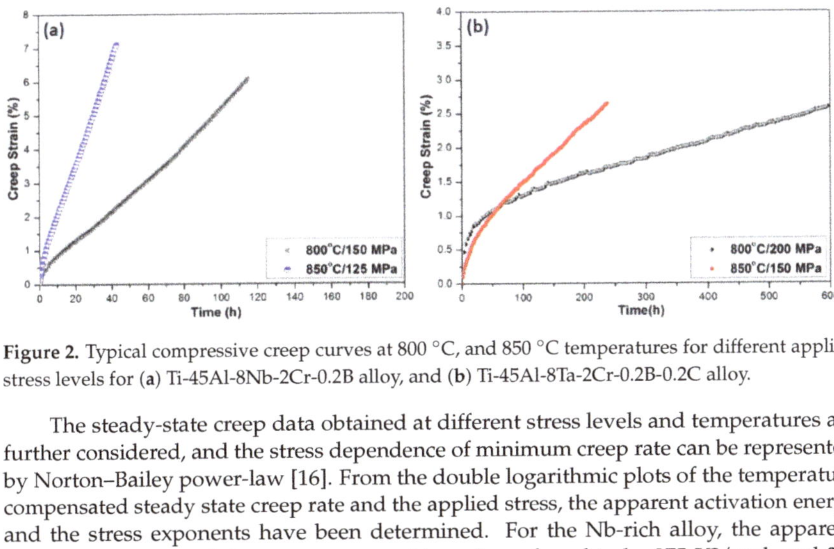

Figure 2. Typical compressive creep curves at 800 °C, and 850 °C temperatures for different applied stress levels for (**a**) Ti-45Al-8Nb-2Cr-0.2B alloy, and (**b**) Ti-45Al-8Ta-2Cr-0.2B-0.2C alloy.

The steady-state creep data obtained at different stress levels and temperatures are further considered, and the stress dependence of minimum creep rate can be represented by Norton–Bailey power-law [16]. From the double logarithmic plots of the temperature compensated steady state creep rate and the applied stress, the apparent activation energy and the stress exponents have been determined. For the Nb-rich alloy, the apparent activation energy and the stress exponent have been found to be 375 KJ/mol, and 3.6, respectively [16]. On the other hand, in the current study, the apparent activation energy of 400 KJ/mol, and the stress exponent of 5.0 have been estimated for the Ta-rich alloy [17]. The value of the activation energy for self-diffusion of Ti and Al in TiAl were reported to be 250–295 kJ/mol and 358 kJ/mol, respectively [18]. The calculated activation energy values of 375–400 kJ/mol for the Nb- and Ta-rich γ-TiAl alloys are higher than that reported for self-diffusion of Ti in TiAl. The higher value of activation energy of creep deformation for the present alloys could be due to the presence of Nb and Ta alloying elements, which are reported to have a low diffusion coefficient in both α (Ti-base solid solution with hexagonal structure) and γ (TiAl) phase. It indicates that Nb and Ta alloying to TiAl alloys substantially increases the activation energy of self-diffusion of Ti in γ-TiAl.

It is to be noted that the stress exponent values obtained for the Nb and Ta containing alloys in the present study are very similar to the stress exponents reported in several TiAl based alloys. In particular, values agree well with that of the Ti-46Al-Ta alloy whose composition is somewhat close to that of the present alloy [19,20]. The higher stress exponent value suggests that creep is controlled by diffusion assisted dislocation climb.

It is clear from the creep curves and corresponding analyses that Ta-rich γ-TiAl based alloy shows superior creep resistance compared to the Nb-rich alloys. At a comparable creep deformation condition, Ta-rich alloy exhibits significantly lower steady state creep rate. Here, we have reported the creep curves for Ta-rich alloys at higher stress so as to produce comparable creep strain for evaluation of deformation microstructure. In the present article we will explore the microstructural features responsible for the difference in creep behavior of Nb- and Ta-rich alloys.

3.3. *Microstructural Characterization of Crept Samples*

Microstructural features of the post-creep samples of both the alloys have been examined using electron microscopy (SEM, EBSD) on the longitudinal sections in terms of degradation of the lamellar structures, dynamic recrystallization, and phase transformation. It has been found that there is no major change in lamellar colony structure, with marginal change in the lamellar spacing and lathe size of the crept samples of both the alloys at lower temperature (800 °C). For the Nb-rich alloy, in addition to lath coarsening and lamellar fragmentations, dynamically recrystallized γ grains are observed in the investigated crept samples. The detailed microstructural characterization of the Nb-rich alloy over the entire creep testing range can be found elsewhere [16]. Microstructural development of crept Ta-rich alloys is significantly different from those of the Nb-rich alloy. In the Ta-rich alloy, copious precipitation along the colony and lath boundaries along with phase transformations could be noticed, as shown in the EBSD section. This part has been discussed in detail elsewhere [17]. The development of microstructural features of crept Nb-rich and Ta-rich alloys pertaining to the present article have been examined in detail through EBSD as described in the following section.

3.3.1. EBSD Mapping

Microstructural development of crept Ti-45Al-8Nb-2Cr-0.2B alloy subjected to creep deformation at 800 °C/150 MPa and 850 °C/125 MPa is shown in Figure 3 in terms of band contrast map and phase distribution map. At lower temperatures, the disintegration of lamellar structures due to γ-lath coarsening and $\beta \rightarrow \gamma$ transformation at the bulky β grains situated on the colony boundary could be noticed (Figure 3a,c). At higher temperature (850 °C), the microstructure is characterized by both $\beta \rightarrow \gamma$ transformation as well as $\beta \rightarrow \alpha_2$ transformations along with γ-lath coarsening and partial fragmentation of laths (Figure 3b,d).

On the other hand, the microstructure of the crept samples of the Ti-45Al-8Ta-2Cr-0.2B-0.2C alloy reveals considerable modification of the initial condition. Figure 4 exhibits typical microstructures of crept samples subjected to creep deformation at 800 °C/200 MPa and 850 °C/150 MPa. It is evident that γ-lath coarsening, $\beta \rightarrow \gamma$ transformation and considerable coarsening of transformed γ-grains at the colony boundaries are common features of the crept microstructure at both the temperatures. At lower temperature, the stability of γ-phase is predominant, as shown in Figure 4c. With increase in creep deformation temperature to 850 °C, in addition to the transformation of remnant β phase to γ, there are α_2 grains at the colony boundary area, possibly due to a $\beta \rightarrow \alpha_2$ transformation (Figure 4d). In contrast to the microstructural development of Nb-rich alloy, the crept Ta-rich alloy samples show precipitation of a Ta-rich τ-phase (Ti_4Al_3Ta, isomorphous with the reported τ-phase based on Ti_4Al_3Nb; the crystal data (CIF file) of Ti_4Al_3Ta for Kikuchi bands was generated from reported phase Ti_4Al_3Nb by replacing the Nb atoms with Ta atoms. The Kikuchi bands of Ti_4Al_3Ta phase were detected in EBSD with a mean angular deviation (MAD) value less than one, which indicates good matching of the simulated patterns. Extent of the τ-phase precipitation increases with the increasing temperature of creep deformation.

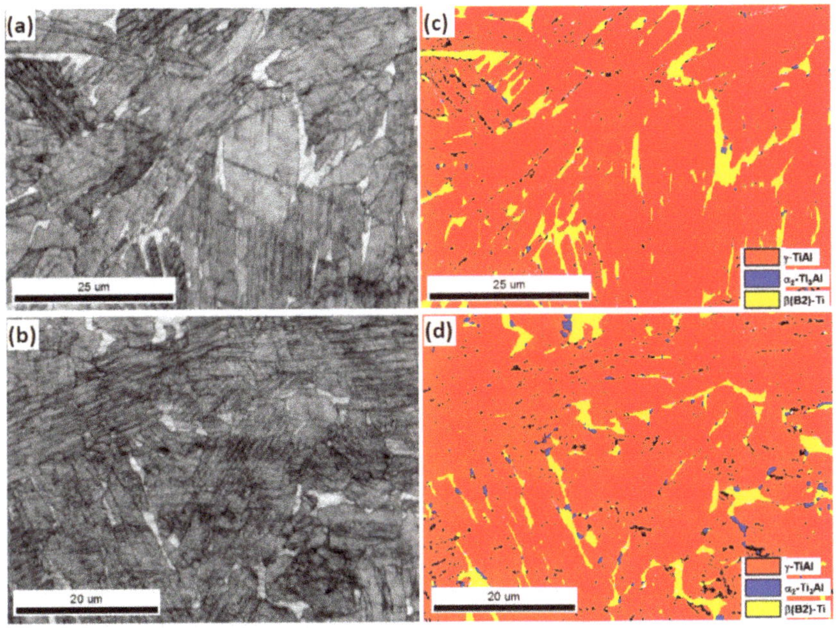

Figure 3. (**a**,**c**) EBSD band contrast map, and corresponding phase map of the Ti-45Al-8Nb-2Cr-0.2B alloy sample subjected to creep deformation at 800 °C/150 MPa. (**b**,**d**) same for the sample subject to creep at 850 °C/125 MPa.

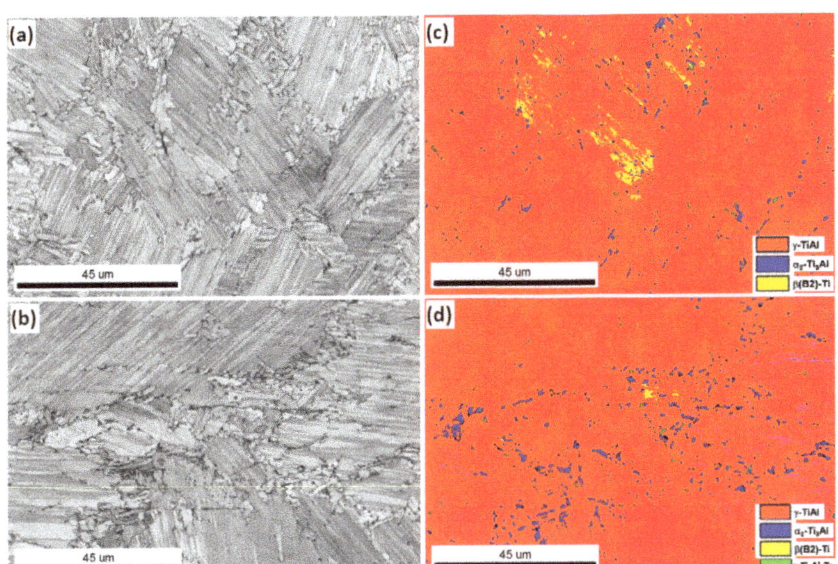

Figure 4. (**a**,**c**) EBSD band contrast map, and corresponding phase map of the Ti-45Al-8Ta-2Cr-0.2B-0.2C alloy sample subjected to the creep deformation at 800 °C/200 MPa. (**b**,**d**) same for the sample subject to creep at 850 °C/150 MPa.

3.3.2. Transmission Electron Microscopy

Figure 5 represents the bright field TEM images showing comparison of microstructural development of Nb- and Ta-rich alloys crept at 800 °C (Figure 5a,c) and 850 °C (Figure 5b,d). It is evident from the TEM images that the lamellar structure is largely retained for the Nb-rich alloy during the creep deformation at 800 °C (Figure 5a), whereas considerable lath coarsening can be observed at higher temperature (Figure 5b). On the other hand, Ta-rich alloy shows considerable lamellar disintegration in terms of fragmentation and thinning of α_2 laths, and formation of recrystallized γ-grains at both the creep deformation temperatures (Figure 5c,d). It could further be noticed that there is lamellar instability accompanied by copious precipitation of phases based on α_2 and τ. The complex co-precipitation of multiple phases makes it difficult to precisely determine their identity. It is, therefore, required that microstructural development be evaluated using an advanced technique such as ACOM with PED for a detailed characterization.

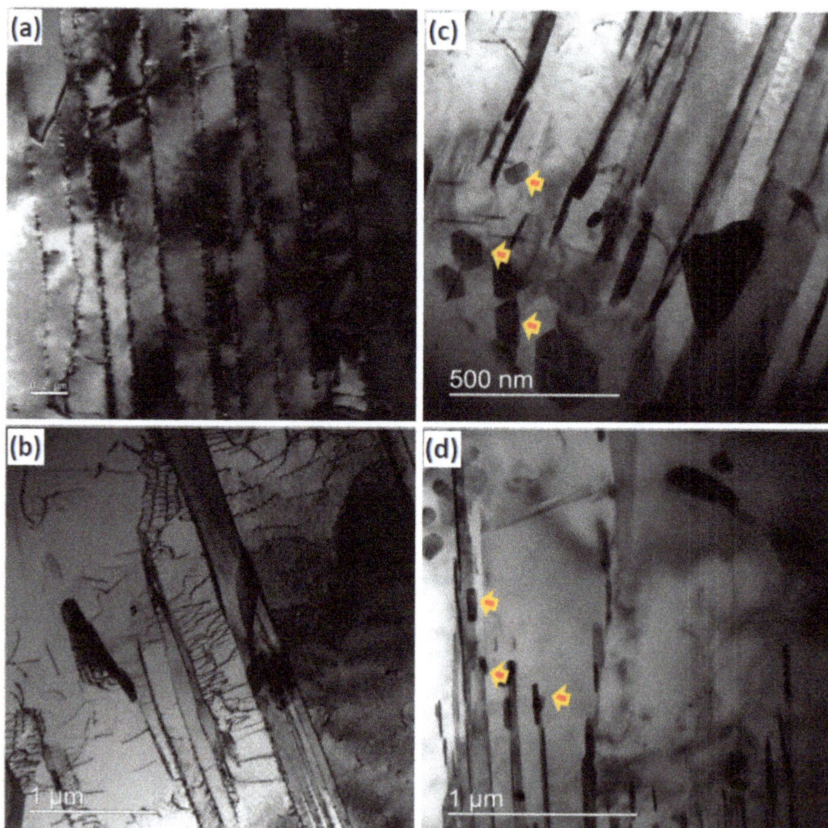

Figure 5. Bright field TEM images showing development of microstructures of the Ti-45Al-8Nb-2Cr-0.2B alloy samples crept at (**a**) 800 °C/150 MPa, and (**b**) 850 °C/125 MPa. (**c**,**d**) Same for the alloy Ti-45Al-8Ta-2Cr-0.2B-0.2C crept at 800 °C/200 MPa, and 850 °C/150 MPa, respectively.

3.3.3. ACOM with PED

The results of ACOM along with PED analysis of the sample crept at 800 °C/200 MPa are displayed in Figure 6, showing the index quality map and corresponding inverse pole figure map (IPF), phase map, and kernel average misorientation (KAM) map. The coarsening of γ-laths and precipitation of τ-phase after creep deformation can clearly be observed from Figure 6a,c. However, coarse α_2-laths still remain, and relatively fine laths tend to dissolve. Further, fine Ti_3Al particles are observed which could possibly form by fragmentation of laths and/or precipitation at the prior inter-lamellar and colony boundaries. The KAM map indicates the spatial localization of strain gradients developed in the microstructure. It could be noticed that the higher strain gradient regions are associated with the precipitation of τ and α_2-particles. The coarse γ-laths are relatively strain free indicating strong dynamic recovery during the creep deformation.

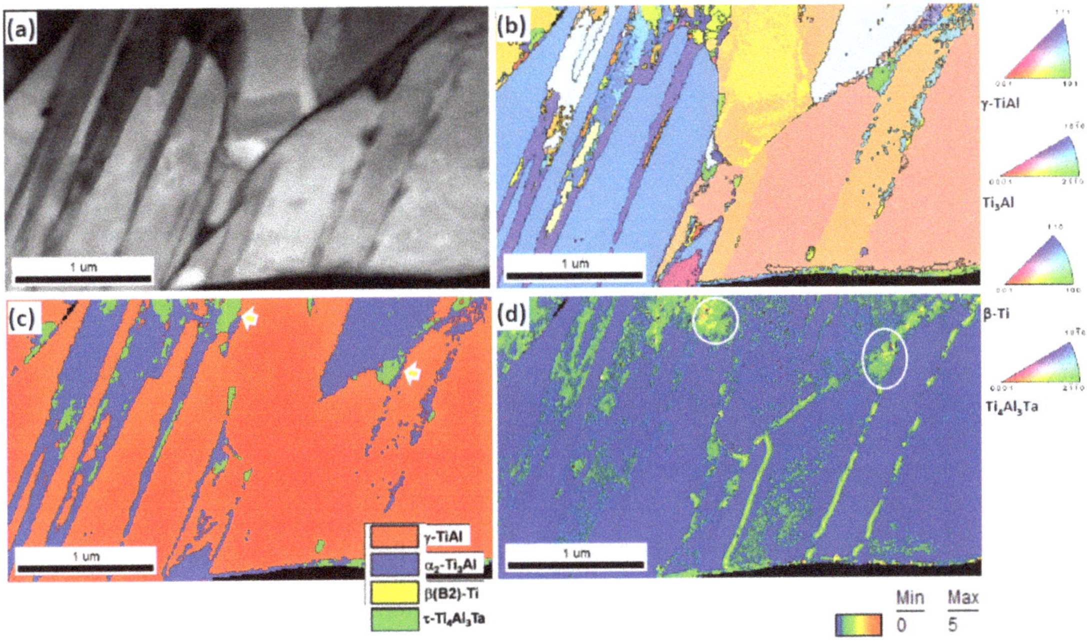

Figure 6. (**a**) Index quality map acquired through ACOM technique of the Ti-45Al-8Ta-2Cr-0.2B-0.2C crept at 800 °C/200 MPa. (**b–d**) Corresponding IPF map, phase map, and KAM map, respectively.

Similar ACOM with PED analyses of the Ta-rich alloy sample crept at 850 °C/150 MPa further corroborate the observations described above. The index quality map and IPF map, phase map, and KAM map of Ta-rich alloy sample are displayed in Figure 7. Considerable coarsening of the γ-laths (an estimate of average γ-lath size of the as-cast material and crept sample are 280 $\pm$ 105 nm, 360 $\pm$ 55 nm, respectively measured from high magnification SEM images) and remnants of the α_2-laths are evident from the phase maps (Figure 7c). The nature of prior α_2-laths can be confirmed from the fact that there are exactly same orientations of the remnant laths as depicted in Figure 7b. It is to be noted that in a $\gamma + \alpha_2$ colony structure the α_2-laths inherit the parent phase orientation from which $\alpha \rightarrow \gamma$ transformation occurs. The precipitation of τ-phase can be noticed along the boundary of a γ-grain and colony boundaries (Figure 7c). Similar to Ta rich sample crept at 800 °C/200 MPa, the KAM map for the sample crept at 850 °C/150 MPa indicates the presence of localized strain gradients near the regions of the precipitation of τ and α_2-particles.

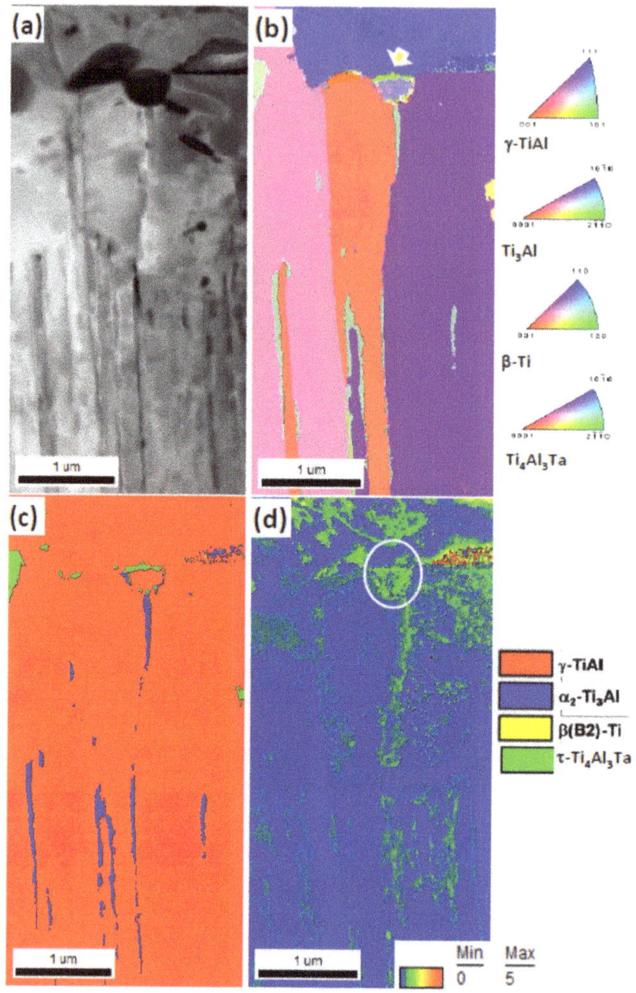

Figure 7. (**a**) Index quality map acquired through ACOM technique of the Ti-45Al-8Ta-2Cr-0.2B-0.2C crept at 850 °C/150 MPa. (**b**–**d**) Corresponding IPF map, phase map, and KAM map, respectively.

One of the interesting aspects of microstructural development of the crept Ta-rich alloy is the precipitation of τ and α_2-particles and the relatively high strain gradient associated with them. It is to be noted that the relatively superior creep resistance of the Ta-rich alloy is rather surprising considering the fact the lamellar structure of the Nb-rich alloy has been shown to be more stable under the creep deformation condition studied here. This indicates the possible role of the precipitation of τ and α_2-particles in improving the creep resistance of the Ta-rich alloy. The existence of higher strain gradients associated with precipitation zones (Figures 6d and 7d) corroborates well with such an interpretation (a few precipitated τ-particles are marked with arrowheads and relevant areas are encircled in KAM maps in Figures 6 and 7). Conventional TEM bright field imaging of the dislocation structures near the precipitation zones further confirms the results. Figure 8 shows the development of dislocation structure of crept Ta-rich alloy. It is evident from the TEM images that extensive dislocation networks are present near the precipitation zones. Considering the typical dislocation structure of coarse γ-laths and dynamically recrystallized γ-grains, such a dense dislocation network around the precipitates indicates "pinning" effect against the dislocation movement, and consequently, dynamic recovery will be hindered.

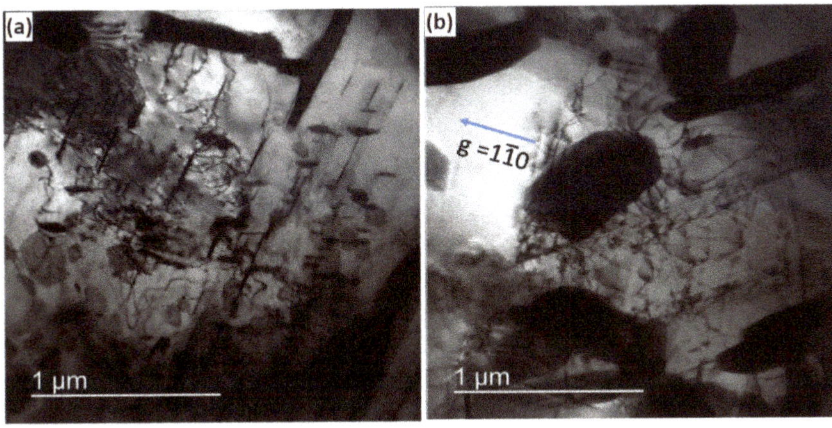

Figure 8. Bright field TEM images showing development of dislocation structures of the Ti-45Al-8Ta-2Cr-0.2B-0.2C crept at (**a**) 800 °C/200 MPa, and (**b**) 850 °C/150 MPa utilizing a near two-beam imaging condition.

From the foregoing discussion, it is evident the precipitation of τ-phase assumes greater significance in the context of improved creep resistance of Ta-rich γ-TiAl based alloys. Hence, transformation of the τ-phase has been examined to a greater extent. The crystallographic nature of the τ-phase reveals that it has $B8_2$ type structure (space group: P63/mmc, Pearson symbol: hP6) and the ternary hexagonal Ti_4Al_3Nb phase has been shown to have a similar symmetry with Ni_2In, as reported by Witusiewicz et al. [21] in Nb-rich Ti-46Al-8Nb alloy. The present investigation also suggests that Ti_4Al_3Ta phase could be considered as isomorphous to the Ti_4Al_3Nb phase with replacement of Ta atoms at the Nb sites. It has been reported in literature that Ti_4Al_3Ta phase is known to form from β (B2) phase [22] through intermediate ω phase. Direct transformation of β(B2) phase to $B8_2$ is not reported in the literature and needs further investigation.

However, Lapin et al. [19,20] have suggested that the τ-phase forms at the expense of the α_2-lathes by partially transforming to the γ-matrix and τ-particles during creep deformation (i.e., $\alpha_2 \rightarrow \gamma + \tau$). In the present study, the τ-phase is often found to be associated with α_2-laths, which tends to support the later view. However, a closer examination of the crystallographic orientation relationship (OR) between the γ, τ, and α_2-phases reveals that there exists a strong OR between γ and τ-phase, whereas the OR between α_2 and τ has a misorientation >15°.

Figure 9 confirms the existence of OR between the γ- and τ-phases in terms of $\langle 111\rangle\gamma//\langle 0001\rangle\tau$; $\langle 110\rangle\gamma//\langle 10\text{–}10\rangle\tau$. However, considering the fact that τ-phase composition (Ti = 45–48%, Al = 41–42%, Ta = 9–10%, Cr = 2–3%) is significantly richer in Ta, it is unlikely that γ-laths (Ti = 46–47%, Al = 45–46%, Ta = 6–7%, Cr = 1–2%) will directly transform to τ-phase. Further considerations of the locations of the τ-phase formation near the remnant β(B2)-phases (Ti = 39–41%, Al = 37–38%, Ta = 13–14%, Cr = 8–9%) in the initial material and almost complete transformation of β(B2)-phases during the creep deformation of Ta-rich alloys suggest that the remnant β(B2)-phases phase transforms simultaneously to α_2-phase and τ-phase by partitioning of Ta. It has been shown in an earlier study [16] that transformation β(B2)-phases present in colony boundaries to γ-grains bears the general OR of $(110)_\beta//(111)\gamma$; $[111]_\beta//[110]\gamma$. Hence, it is likely that the transformed γ-phase will have an OR with τ-phase even though τ-phase does not directly transform from γ-phase. It is, therefore, postulated that the evolution of a hexagonal τ-phase could possibly occur through the transformation sequence of $\beta(B2) \rightarrow \alpha_2 + \tau$. Further micro-chemical analyses in conjunction with the high resolution EBSD of the Ta-rich sample crept at 800 °C/200 MPa illustrate the compositional characteristics of the possible transformation products (i.e., α_2, τ), as summarized in Figure 10. It is evident from the IPF and band contrast maps (Figure 10a,b) that the transformed α_2, τ-phases are noticed mainly at the recrystallized γ-grain boundaries. While the precipitated τ-phases have a characteristic narrow composition (Ti = 46.1%, Al = 41.5%, Ta = 10.5%, Cr = 2.3%), the α_2-phase shows an unusual scatter in composition (Ti = 37–49%, Al = 36–41%, Ta = 7–18%, Cr = 2–8%). It is rather interesting to note that the thin, irregular shaped α_2-phases often associated with τ-particles show highly Ta-rich composition (Ti = 40.5%, Al = 40.4%, Ta = 17.3%, Cr = 1.8%), as shown in Figure 10e. On the other hand, course and equiaxed α_2-grains as well as the remnant α_2-laths have a more consistent composition of Ti = 48–49%, Al = 42–43%, Ta = 6–7%, Cr = 1–2%. Such a unique compositional characteristics of the α_2-phases differentiates the transformed products with rather high Ta-rich composition and further corroborates the postulated transformation from the Ta-rich parent β-phase. A careful examination of the transformed regions reveals the frequent presence of very thin peripheral layer of β-phase around the α_2 supplementing the proposed transformation path.

ACOM with PED analyses have been further extended to the Ti-45Al-8Nb-2Cr-0.2B alloy sample subjected to creep at 800 °C/150 MPa for comparison. Figure 11 summarizes the key microstructural findings. Compared to the Ta-rich crept samples, the β(B2)-phase of the Nb-rich samples appears to be more stable under the creep deformation conditions. As depicted in Figure 11c, τ-phase and α_2-phase precipitation have hardly been observed in the microstructure. Only localized thin layers of α_2-phase could be noticed. More importantly, relatively lower and inconsistent strain gradient could be detected near the α_2-phase from the KAM map (Figure 11d). It clearly points out that, unlike the case of the precipitation of τ and α_2-particles in Ta-rich alloys, the transformed α_2-phase has hardly any influence on the creep behavior of the Nb-rich alloy.

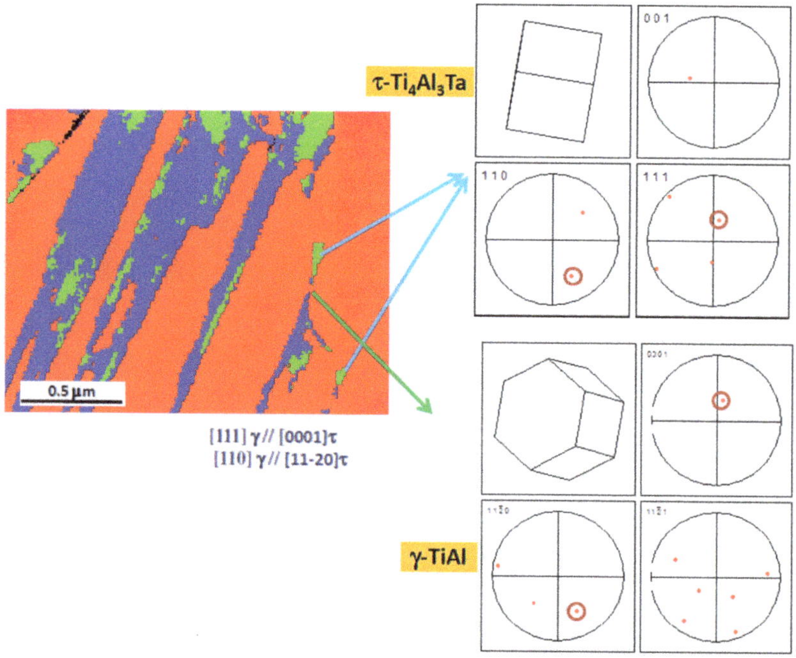

Figure 9. Orientation relationship (OR) between the τ-phase and surrounding γ-matrix of the Ti-45Al-8Ta-2Cr-0.2B-0.2C crept at 800 °C/200 MPa.

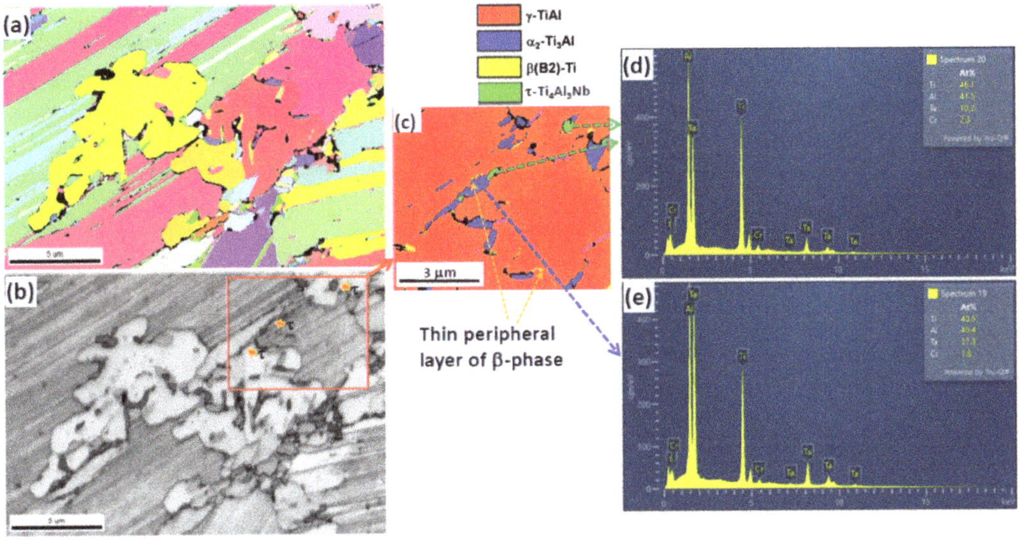

Figure 10. (**a**) EBSD-IPF map, and (**b**) BC map of the Ti-45Al-8Ta-2Cr-0.2B-0.2C crept at 800 °C/200 MPa. (**c**) Selected portion of the phase map (marked region of (**b**)) shows the coexistence of α_2 and τ-particles as transformation products. (**d**,**e**) Corresponding EDS patterns of α_2 and τ-particles, respectively.

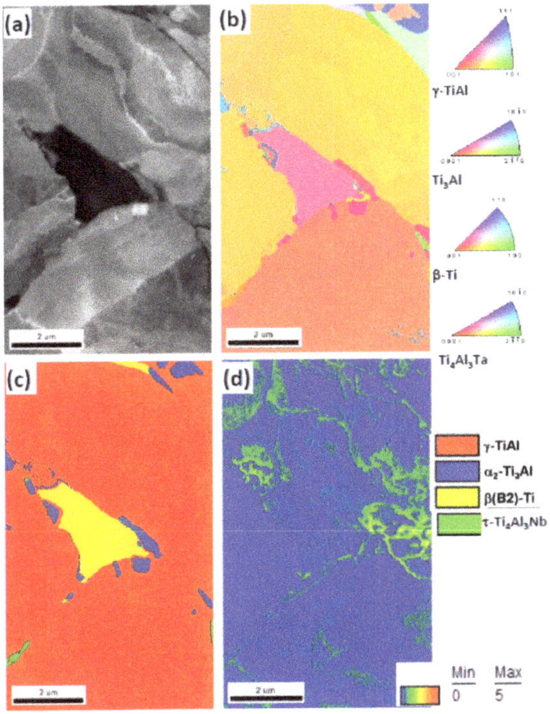

Figure 11. (**a**) Index quality map acquired through ACOM technique of the Ti-45Al-8Nb-2Cr-0.2B crept at 800 °C/150 MPa. (**b**–**d**) Corresponding IPF map, phase map, and KAM map, respectively.

4. Conclusions

Compressive creep behavior of two fourth generation γ-TiAl alloys, Ti-45Al-8Nb-2Cr-0.2B and Ti-45Al-8Ta-2Cr-0.2C-0.2B has been evaluated and correlated with microstructure of the alloys. Followings are the major outcome of the present study.

(a) Compressive creep curves of the alloys studied under a range of deformation conditions (800–850 °C, 125–200 MPa) exhibit typical primary regime followed by a well-defined steady state creep regime. The stress dependence of the steady state creep rates can be described by the Norton–Bailey power-law with stress exponent being in the range 3.6 to 4.5 and apparent activation energy in the range 375 to 398 kJ/mol. The observed values of stress exponent and activation energy suggest that the creep in the alloy is controlled by diffusion which is assisted by dislocation climb.

(b) Detailed microstructural comparison of the pre- and post-creep samples reveals that compared to the Nb-rich alloy the Ta-rich alloy exhibit higher tendency of lamellar structure disintegration in terms of γ-lath coarsening and fragmentation of α_2-laths in the creep deformation range studied here. Microstructure of the crept Ta-rich alloy is further decorated with precipitation of multiple phases. In addition to the α_2-precipitates, τ-phase (Ti_4Al_3Ta), precipitation occurs during creep along the lath and the colony boundaries. Precipitation of τ-phase (Ti_4Al_3Nb) corresponding to the Nb-rich alloy is scarce in the crept microstructure.

(c) It has been observed that a definite OR exists between the precipitated τ-phase and its surrounding matrix γ-phase in the form of $\langle 111\rangle\gamma//\langle 0001\rangle\tau$; $\langle 110\rangle\gamma//\langle 10\text{–}10\rangle\tau$. However, no such OR could be found between α_2 and τ. This implies that the evolution of a hexagonal τ-phase could possibly occur through the transformation sequence of $\beta(B2) \rightarrow \alpha_2 + \tau$.

(d) In spite of a tendency for lamellar structure instability of the Ta-rich alloy, the creep resistance of the Ta-rich alloy is superior to that of the Nb-rich alloy. The microstructural attribute to such behavior has been found to be the dislocation pinning of precipitated second phases, in particular the τ-phase (Ti_4Al_3Ta).

Author Contributions: Conceptualization, C.M.; investigation, methodology, formal analysis, data curation, V.S., C.M., S.R., C.M.O. and R.S.; writing—original draft preparation, V.S. and C.M.; supervision, C.M. and P.G.; funding acquisition, P.G. All authors have read and agreed to the published version of the manuscript.

Funding: This research was funded by Defence Research and Development Organization (DRDO), Government of India.

Data Availability Statement: The data will be made available on reasonable request.

Acknowledgments: The authors are also grateful to Deepak Kumar, Kaling Sahoo, DMRL for help in carrying out experiments.

Conflicts of Interest: The authors declare no conflict of interest. The funders had no role in the design of the study; in the collection, analyses, or interpretation of data; in the writing of the manuscript, or in the decision to publish the results.

References

1. Appel, F.; Paul, J.D.H.; Oehring, M. *Gamma Titanium Aluminide Alloys: Science and Technology*; Wiley-VCH Verlag GmbH & Co: Weinheim, Germany, 2011.
2. Kardashova, S.I.; Yu, A.; Razumovskii, L.; Razumovskii, I.M. Diffusion coarsening of the lamellar structure in two-phase Ti-47.5 at.% Al intermetallic alloy. *Acta Metall. Mater.* **1994**, *42*, 3341–3348. [CrossRef]
3. Bartholomeusz, M.F.; Wert, J.A. The effect of thermal exposure on microstructural stability and creep resistance of a two-phase TiAl/Ti3Al lamellar alloy. *Metall. Mater. Trans. A* **1994**, *25*, 2371–2381. [CrossRef]
4. Huang, S.C. *Alloying Consideration in Gamma-Based Alloys, Structural Intermetallics*; Darolia, R., Lewandowski, J.J., Liu, C.T., Martin, P.L., Miracle, D.B., Nathal, M.V., Eds.; TMS: Warrendale, PA, USA, 1993; pp. 299–307.
5. Xiang, L.L.; Zhao, L.L.; Wang, Y.L.; Zhang, L.Q.; Lin, J.P. Synergistic effect of Y and Nb on the high temperature oxidation resistance of high Nb containing TiAl alloys. *Intermetallics* **2012**, *27*, 6–13. [CrossRef]

6. Sienkiewicz, J.; Kuroda, S.; Murakami, H.; Araki, H.; Gizynski, M.; Kurzydlowski, K.J. Microstructure and Oxidation Performance of TiAl-(Cr,Nb,Ta) Coatings Fabricated by Warm Spray and High-Velocity Oxy-Fuel Spraying. *J. Thermal. Spray Tech.* **2019**, *28*, 563–579. [CrossRef]
7. Singh, V.; Kumar, A.; Mondal, C.; Bhattacharjee, P.P.; Ghosal, P. Hot deformation of high-Nb-containing γ-TiAl alloy in the temperature range of 1000–1200 °C: Microstructural attributes to hot workability. *SN Appl. Sci.* **2019**, *1*, 366. [CrossRef]
8. Lapin, J.; Pelachová, T.; Dománková, M. Creep behaviour of a new air-hardenable intermetallic Ti-46Al-8Ta alloy. *Intermetallics* **2011**, *19*, 814–819. [CrossRef]
9. Clemens, H.; Smarsly, W. Light-weight intermetallic Titanium aluminides-status of research and development. *Adv. Mater. Res.* **2011**, *278*, 551–556. [CrossRef]
10. Imayev, V.; Imayev, R.; Khismatullin, T.; Oleneva, T.; Gühter, V.; Fecht, H.-J. Microstructure and processing ability of β-solidifying TNM-based γ-TiAl alloys. *Mater. Sci. Forum* **2010**, *638–642*, 235–240. [CrossRef]
11. Wang, J.G.; Nieh, T.G. Creep of a beta phase-containing TiAl alloy. *Intermetallics* **2000**, *8*, 737–748. [CrossRef]
12. Gabrisch, H.; Stark, A.; Schimansky, F.-P.; Wang, L.; Schell, N.; Lorenz, U.; Pyczak, F. Investigation of carbides in Ti–45Al–5Nb–xC alloys ($0 \leq x \leq 1$) by transmission electron microscopy and high energy-XRD. *Intermetallics* **2013**, *33*, 44–53. [CrossRef]
13. Available online: https://www.ebsd.com/ois-ebsd-system/pseudosymmetry-correction (accessed on 23 November 2021).
14. Rauch, E.F.; Véron, M.; Portillo, J.; Bultreys, D.; Maniette, Y.; Nicolopoulos, S. Automatic Crystal Orientation and Phase Mapping in TEM by Precession Diffraction. *Microsc. Anal.* **2008**, *22*, s5–s8.
15. Singh, V.; Mondal, C.; Bhattacharjee, P.P.; Ghosal, P. Microstructural Characterization by Automated Crystal Orientation and Phase Mapping by Precession Electron Diffraction in TEM: Application to Hot Deformation of a γ-TiAl based Alloy. *Microsc. Micanalysis* **2019**, *25*, 1457–1465. [CrossRef] [PubMed]
16. Singh, V.; Mondal, C.; Sarkar, R.; Bhattacharjee, P.P.; Ghosal, P. Compressive creep behavior of a γ-TiAl based Ti–45Al–8Nb–2Cr-0.2B alloy: The role of β(B2)-phase and concurrent phase transformations. *Mat. Sci. Eng. A* **2020**, *774*, 138891. [CrossRef]
17. Singh, V.; Sahoo, K.; Mondal, C.; Satyanarayana, D.V.V.; Ghosal, P. *High Temperature Compressive Creep Properties of Ti–45Al–8Ta–2Cr-0.2B-0.2C Alloy for Aerospace Applications*; DMRL Technical Report, DMRL-EMG-234-2020; DMRL: Hyderabad, India, October 2020.
18. Herzig, C.H.; Przeorski, T.; Mishin, Y. Self-diffusion in γ-TiAl: An experimental study and atomistic calculations. *Intermetallics* **1999**, *7*, 389–404. [CrossRef]
19. Lapin, J.; Pelachová, T.; Witusiewicz, V.T.; Dobročka, E. Effect of long-term ageing on microstructure stability and lattice parameters of coexisting phases in intermetallic Ti–46Al–8Ta alloy. *Intermetallics* **2011**, *19*, 121–124. [CrossRef]
20. Lapin, J.; Gabalcová, Z.; Pelachová, T.; Bajana, O. Microstructure and mechanical properties of a cast intermetallic Ti-46Al-8Ta alloy. *Mater. Sci. Forum* **2010**, *638–642*, 54–60. [CrossRef]
21. Witusiewicz, V.T.; Bondar, A.A.; Hecht, U.; Velikanova, T.Y. The Al–B–Nb–Ti system: IV. Experimental study and thermodynamic re-evaluation of the binary Al–Nb and ternary Al–Nb–Ti systems. *J. Alloys Compd.* **2009**, *472*, 133–161. [CrossRef]
22. Bendersky, L.A.; Boettinger, W.J.; Bruton, B.P.; Biancaniello, F.S. The formation of ordered ω-Ordered related phases in alloys of composition Ti_4Al_3Nb. *Acta Metal. Mater.* **1990**, *38*, 931–943. [CrossRef]

symmetry

MDPI

Article

Structure Determination Feasibility of Three-Dimensional Electron Diffraction in Case of Limited Data

Partha Pratim Das [1,†], Sergi Plana-Ruiz [1,2,†], Athanassios S. Galanis [1], Andrew Stewart [3,*], Fotini Karavasili [4], Stavros Nicolopoulos [1,*], Holger Putz [5], Irene Margiolaki [4], Maria Calamiotou [6] and Gianluca Iezzi [7,8]

1 NanoMEGAS SPRL, Rue Èmile Claus 49 bte 9, 1050 Brussels, Belgium
2 Servei de Recursos Científics i Tècnics, Universitat Rovira i Virgili, Avinguda dels Països Catalans 26, 43007 Tarragona, Catalonia, Spain
3 Department of Chemistry, University College London, London WC1H 0AJ, UK
4 Section of Genetics, Cell Biology and Development, Department of Biology, University of Patras, GR-26500 Patras, Greece
5 CRYSTAL IMPACT, Dr. H. Putz & Dr. K. Brandenburg GbR, 53227 Bonn, Germany
6 Section of Condensed Matter Physics, Physics Department, University of Athens, GR-15784 Athens, Greece
7 Dipartimento di Ingegneria & Geologia (INGEO), University G. d'Annunzio of Chieti-Pescara, Via Dei Vestini 30, 66100 Chieti, Italy
8 INGV (Istituto Nazionale di Geofisica e Vulcanologia) Roma, Via di Vigna Murata 605, 00143 Roma, Italy
* Correspondence: andy.stewart@ucl.ac.uk (A.S.); info@nanomegas.com (S.N.)
† These authors contributed equally to this work.

Abstract: During the last two decades, three-dimensional electron diffraction (3D ED) has undergone a renaissance, starting with the introduction of precession (Precession Electron Diffraction Tomography, PEDT) that led to variations on the idea of collecting as much of the diffraction space as possible in order to solve crystal structures from sub-micron sized crystals. The most popular of these acquisition methods is based on the continuous tilting/rotation of the crystal (so-called Microcrystal Electron Diffraction, MicroED) akin to the oscillating crystal method in X-ray crystallography, which was enabled by the increase of sensitivity and acquisition speed in electron detectors. While 3D ED data is more complex than the equivalent X-ray data due to the higher proportion of dynamical scattering, the same basic principles of what is required in terms of data quality and quantity in order to solve a crystal structure apply; high completeness, high data resolution and good signal-to-noise statistics on measured reflection intensities. However, it may not always be possible to collect data in these optimum conditions, the most common limitations being the tilt range of the goniometer stage, often due to a small pole piece gap or the use of a non-tomography holder, or the position of the sample on the TEM grid, which may be too close to a grid bar and then the specimen of interest becomes occluded during tilting. Other factors that can limit the quality of the acquired data include the limited dynamic range of the detector, which can result on truncated intensities, or the sensitivity of the crystal to the electron beam, whereby the crystallinity of the particle is changing under the illumination of the beam. This limits the quality and quantity of the measured intensities and makes structure analysis of such data challenging. Under these circumstances, traditional approaches may fail to elucidate crystal structures, and global optimization methods may be used here as an alternative powerful tool. In this context, this work presents a systematic study on the application of a global optimization method to crystal structure determination from 3D ED data. The results are compared with known structure models and crystal phases obtained from traditional ab initio structure solution methods demonstrating how this strategy can be reliably applied to the analysis of partially complete 3D ED data.

Keywords: 3D ED; MicroED; precession electron diffraction; crystal structure determination; global optimization methods

1. Introduction

Structure solution of crystalline materials via electron diffraction (ED) methods has a long history stretching back several decades [1–4]. While the technique has an obvious appeal of being able to solve structures from the smallest of crystals, it was not widely adopted owing to major challenges with data collection and analysis. In fact, it was initially believed that only zone axis diffraction patterns with the corresponding image was the only method to overcome the interpretation of highly dynamical zone axis data, a very slow and technically challenging process [5]. The inherent dynamical diffraction effects present in ED data are due to the $10^{3–4}$ stronger interactions of electrons with respect to X-rays, a feature that allows to focus an electron beam to few nanometres but increases the likelihood of multiple scattering events [6–8].

Direct methods were one of the initially available algorithms for crystal structure determination that were developed for X-ray diffraction data. This means that the kinematic theory of diffraction was assumed to process the reflection intensities, i.e., a single scattering event for each photon interacting with a crystal. Nevertheless, direct methods are based on statistical distributions that makes them more robust to noise than other approaches, such as charge-flipping, which uses the measured intensities directly to reconstruct the crystal phase. This can be seen in the fact that direct methods were developed in an era when intensities were being measured by eye and categorised into a series of intensity bins, from strong to weak. However, ED patterns only from zone axes are too dynamical even for the robustness of direct methods [9].

The key advantage of structure analysis by using a transmission electron microscope (TEM) is the ability to probe and extract information from nanometre-sized volumes of individual crystals/materials, which is often challenging or impossible via traditional single-crystal X-ray diffraction methods available in most structural analysis laboratories. It was not until the development of precession electron diffraction (PED) by Vincent and Midgley [10], and Own [11], as well as the commercial availability of a precession-signal generator (SpinningSTAR, DigiSTAR from NanoMEGAS SPRL) [12,13], that ED took a major step forward on the analysis of nanocrystals [12]. PED leads to higher number of reflections in a diffraction pattern, better integration of reflection intensities and reduced dynamical effects. Besides structure solution, PED has also been proved to be advantageous for phase and orientation mapping [14], strain analysis [15] and electron pair distribution study of nanoparticles [16].

PED demonstrated that collecting data away from the perfectly aligned zone axis was advantageous by significantly reducing the dynamical diffraction effects due to the integration of the reflection intensities [10]. This initial idea was then extended by Ute Kolb's group, which introduced the sequential acquisition of PED patterns from a random crystal orientation without the need to find the different zone-axes, further reducing the dynamical effects, and obtaining a quasi-kinematic ED dataset that enabled direct methods to be robust enough to solve structures ab initio. The data acquisition was achieved by combining PED and the acquisition methodology of image tomography, demonstrating for the first time an efficient method for collecting three-dimensional electron diffraction (3D ED) data, an acquisition routine also known as Precession Electron Diffraction Tomography (PEDT) [17–21]. This is performed by tilting the crystal in 1° increments around the tilt axis of the TEM goniometric stage and recording a diffraction pattern with a precessed beam for each tilt. A precession angle of 1° is usually used to ensure that the diffraction space is integrated ~ $\pm 1°$ around the stage position, integrating the Bragg reflections on an electron detector in a manner analogous to the X-ray oscillating crystal method (Figure 1). Data processing is then followed by different available software, e.g., ADT3D [17,18] distributed by Nanomegas SPRL [13], RED [22,23] or PETS2 [24], that have been specifically developed to determine unit cell parameters and extract the recorded reflection intensities from 3D ED data. The extracted intensities are subsequently used for structure solution using either direct methods as implemented in SIR software [25–27] or charge-flipping as implemented in Jana 2006 [28,29]. In recent years, the development of CMOS-based detectors for indirect (optical fibre coupled) and direct electron detection [30,31] have enabled the collection of ED data by continuously rotating the crystal [32,33], a method known as MicroED (Microcrystal Electron Diffraction) [34], thus increasing the easiness to acquire such kind of datasets. With the rapid advances of 3D ED methods, many X-ray crystallography software packages (e.g., XDS, DIALS) [35,36] have also been adapted to process 3D ED data, further lowering the barrier for X-ray crystallographers to familiarise with ED techniques.

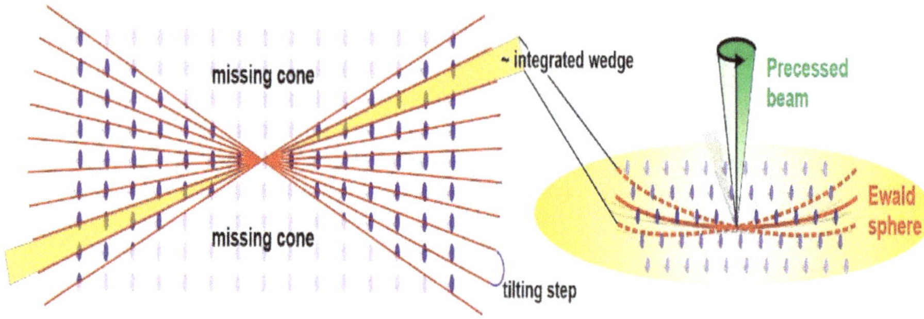

Figure 1. Illustration of the data collection procedure viewed along the tilt axis (**left**) and the use of beam precession to integrate reflections along the excitation error (**right**).

Direct methods and charge-flipping require high data completeness for successful and reliable structure determination [37–40]. The continuous rotation approach depends on the accessibility of modern CMOS-based detectors as well as holders and stages of large tilt range capabilities. Direct electron cameras like hybrid-pixel detectors are more sensitive and can detect electrons more efficiently than charge-coupled devices (CCD's). In addition, hybrid-pixel detectors have a significantly faster readout mechanism, which enables the collection of the continually rotating crystal with minimal to no loss of data in the recorded integrated volume. The higher sensitivity of this technology enables beam sensitive materials to be collected at very low dose conditions that would not be possible with a CCD [32]. In circumstances where only a CCD is available, acquiring sufficient high-quality diffraction data from beam sensitive materials at room temperature is challenging or, in some cases, impossible to allow a successful ab initio crystal structure determination. In this situation of low data completeness and/or low resolution, an alternative approach for crystal structure determination is to apply global optimization methods, since they have been shown to be a particular powerful tool for the retrieval of crystal phases from powder X-ray diffraction data [41,42].

Global optimization is a numerical analysis method attempting to find a global minimum or maximum of a so-called "cost function" (which in turn can be a combination of functions on a given set of variables, here the crystal structure data). Global optimization methods have been extensively used for solving molecular and ionic structures from powder X-ray diffraction [26,42–51] as well as from ED data [52–54]. This approach requires generating initial structure models either at random or based on the known molecular connectivity for organics or known polyhedral geometry for inorganic compounds [55–57]. In global optimization methods, the positions of individual atoms as well as the global shift and rotation of the molecule and non-constrained torsional angles are free to vary until an optimal agreement between the observed and calculated reflection intensities is obtained. Depending on the complexity of the problem, different global optimization techniques such as Monte Carlo, simulated annealing or genetic algorithms can be employed for structure analysis [58].

In this work, we present the study of 3D ED data of limited quality and resolution from four different inorganic crystalline materials to evaluate the feasibility of successful crystal structure determination. First, the traditional ab initio approaches of direct methods and charge-flipping are used to find and understand the limitations of low-quality data.

Then, the obtained structure models are compared to the ones determined from the global optimization method to show how it can be used as an additional and helpful tool for crystal structure analysis when only limited or constrained data is available.

2. Materials and Methods

2.1. Materials

Four different known inorganic materials have been studied by means of 3D ED: $PrBa_2Cu_3O_7$, $BaCuO_2$, $LiAl_{0.8}Fe_{0.2}(SiO_3)_2$ and $LiAl(SiO_3)_2$. $PrBa_2Cu_3O_7$ is the well-known Pr123 material that lacks the superconductivity properties of the Y123-type materials [59,60], while $BaCuO_2$ is always present as an impurity in the Pr123 samples due to antisite defects (Ba atom on Pr site for Pr123 compound), which has also been previously detected by powder diffraction [60,61]. $LiAl(SiO_3)_2$ (spodumene) is a pyroxene mineral, a single-chain silicate (inosilicate), that is an important source of lithium for ceramics, mobile phones and automotive batteries [62]. $LiAl_{0.8}Fe_{0.2}(SiO_3)_2$ is also an inosilicate from the spodumene family in which the aluminum position has been partially occupied by iron [63–65]. From a crystallographic point of view, Pr123 grows in an orthorhombic crystal system with space group *Pmmm* and lattice parameters a = 3.8605 Å, b = 3.9243 Å and c = 11.6993 Å [60,66]. $BaCuO_2$ is a cubic crystal in the $I\,m\bar{3}m$ space group with a = 18.2700 Å [67], and both lithium-ion compounds are monoclinic crystals with space group $C\,2/c$ and the following lattice parameters: a = 9.474 Å, b = 8.390 Å, c = 5.219 Å and β = 110.07° for $LiAl(SiO_3)_2$ [60,62], and a = 9.490 Å, b = 8.426 Å, c = 5.232 Å and β = 110.14° for $LiAl_{0.8}Fe_{0.2}(SiO_3)_2$ [65].

2.2. 3D ED Acquisitions

Samples for TEM measurements were prepared by crushing the raw material with a mortar and pestle to a fine powder. Then, the specimen was dispersed on a 200 mesh Cu-TEM grid with a continuous amorphous carbon film.

3D ED data were collected with a JEOL 2100 LaB_6 TEM operated at 200 kV in TEM mode with a standard single-tilt JEOL holder, and equipped with a precession signal generator (SpinningStar, provided by Nanomegas) [13]. A side-entry Gatan ES500W camera was used for the acquisition of the diffraction data, which is a CCD detector that views a fluorescent screen tilted 45° with respect to the horizontal plane of the TEM. This detector was one of the first CCD's available for the TEM market and has a low dynamic range (12-bit) as well as low sensitivity compared to more modern CCDs. However, it has been proved that such kind of experimental setup can be used for structure solution and even structure refinements based on the kinematic and even dynamical theory of diffraction in some cases [68,69]. Nevertheless, the dataset of $LiAl(SiO_3)_2$ was acquired with a post-column Gatan UltraScan1000 CCD camera (16-bit) placed on a FEI Tecnai F30 TEM operated at 300 kV in STEM mode for quality comparisons.

Diffraction data were acquired in selected-area electron diffraction mode using a 50-μm selected-area aperture, a 70-μm condenser aperture, and a quasi-parallel beam obtained by spreading the beam across a large area [20,70]. Crystals of around 400 nm in size were selected in imaging mode and the thin edges of the crystals were utilised. Although automatic procedures for this kind of experiments are available nowadays [69,71], diffraction data was collected following a manual procedure. Each targeted crystal was standing at an arbitrary crystallographic axis orientation and diffraction patterns were acquired by tilting the holder between −30° and +30° with a tilt step of 1° (maximum available tilt range according to the used pole pieces and single-tilt holders), consequently recording 61 diffraction patterns. Each diffraction pattern was recorded with a precession angle of 1°, the exception being $BaCuO_2$, where precession was not used to compare results between PED and non-PED data. The tilting of the goniometric stage was controlled by a Digital Micrograph script to ensure that each tilt step corresponded to 1°, which included a feedback loop in the form of a *while* and *if* statements to ensure that the requested tilt angle was obtained with an error of less than 0.05°. Crystal tracking was achieved by defocusing the central beam of the diffraction pattern, whereby it is possible to visualize the crystal as a shadow image inside the zero order (non-diffracted) beam of the pattern, correcting the crystal movement by the x and y shifts of the stage, and refocusing it to collect the subsequent precessed diffraction pattern.

The ADT3D software (also called eADT [13,72,73]) was used to process the acquired dataset. This software merges the recorded diffraction patterns into a 3D reconstruction of the diffraction space accounting for pattern shifts that occur during the acquisition process. The software then determines the unit cell parameters, indexes the reflections, and extracts the intensity of the indexed reflections to produce a *hkl* file ready for crystal structure determination. Datasets with precession were also processed with PETS2 in order to analyse the double-peaked rocking curves caused by the reflection integration of the precession movement [24].

2.3. Structure Determination Tools

Direct Methods & Charge-Flipping

Ab initio crystal structure solutions were obtained via direct methods and the charge-flipping procedure with the *hkl* files obtained from ADT3D and PETS2. Direct methods from the SIR2014 software package with the appropriate ED settings and the charge-flipping implemented in SUPERFLIP inside the Jana 2006 software suite were used to solve the crystal structures [28,74–76].

2.4. Global Optimization Method

In contrast to direct methods, a reciprocal space method, and charge-flipping, an iterative dual space method, the global optimization method typically uses the so-called "direct-space" approach [45–49]. This means that a complete structure model (including unit cell parameters, space group and atomic positions) is proposed at first independently of the diffraction information [46–48,77,78]. Afterwards, this model is repeatedly varied and evaluated by comparing its calculated diffraction pattern to the experimental one. The goal is to reduce the so-called "cost function" (e.g., difference between calculated and experimental diffraction pattern) as far as possible until the global minimum is reached (in the optimal case). If no other constraints are introduced, this straightforward prescription, also known as the "Reverse Monte Carlo"-method [79,80], is limited in practice by the fact that for most crystalline solids the system is quickly trapped in a minimum that does not correspond to a physically reasonable atomic arrangement. The reason for this lies in the landscape of the cost function which includes numerous deep local minima [80]. To resolve (or at least reduce) this problem, prior chemically reasonable knowledge about the crystal structures, like sensible interatomic distances, connectivity (molecules) or atomic charges (ions), can be readily applied with the direct-space approach during the structure solution calculation. For example, in case of molecular structures, the structure solution calculation can be reduced to the determination of the 3-dimensional rotation and bond angles as well as the location of the molecule within the unit cell.

In this study we use the global optimization method as it is implemented in the Endeavour software [50,51]. Endeavour solves crystal structures by applying a combined global optimization of the differences between the calculated and the measured diffraction patterns (evaluated by the R factor) and, at the same time, the potential energy of the system ("Pareto-optimization" [81]) to avoid the aforementioned traps.

Each part of the cost function (potential energy and pattern difference) depends on all atomic coordinates, thus setting up a high dimensional hypersurface. "Merging" both hypersurfaces weakens or even eliminates the minima that belong to only one of the two surfaces and strengthens those which belong to both. Therefore, a sufficiently long global optimization run should sooner or later reach the global minimum of the system corresponding to the correct crystal structure.

In principle, any method for the calculation of the potential energy can be used, but the global optimization often makes ab initio energy calculations not feasible. Ab initio energy calculations have two drawbacks in this context. First, they are relatively time-consuming compared to energy calculations using empirically derived potential functions, while typical Endeavour runs require millions of energy calculations in a reasonable amount of time. Second, although they work quite well in equilibrium states, there is always a certain risk that no convergence can be reached for chemically/physically "unreasonable" atomic arrangements where the common wave functions and approximations may no longer be applicable. Both drawbacks are not favorable for a routine application of the method, especially if non-expert users are involved. Fortunately, it turns out that at least for inorganic compounds that can be described with ions, a combination of the corresponding Coulomb forces [82] and a simple repulsion potential parametrized by minimum interatomic distances is generally sufficient.

In order to find the global minimum of the hypersurface, Endeavour uses "simulated annealing", a Monte Carlo-based global optimization method [83]. The idea of "simulated annealing" is to randomly walk around the hypersurface of the cost function(s), controlled by an algorithm that does not only walk "downhill" but also accepts modifications that cause an increase of the cost function with a certain probability (controlled by the "temperature" parameter). In the beginning of the optimization (structure solution calculation), the "temperature" value (i.e., the control parameter) is rather high, so that the probability that "bad" modifications are accepted is large and the system can walk around and visit large parts of the hypersurface of the cost function(s). When the calculation progresses, the "temperature" is reduced more and more, so that the probability to accept "bad" modifications is decreased. Therefore, the system will most probably be getting trapped in the deepest minimum (with regard to the number of states), which is supposed to be coincident with the deepest (hopefully global) minimum of the hypersurface.

In practice, it turns out that despite the inclusion of the potential energy during the structure solution process the hypersurface created by the two cost function contributions (R factor and potential energy) can still be rough and difficult to optimize. This is especially true if space groups with many different special positions and/or high multiplicities of the general or special positions are used, or diffraction data is of low quality (beam damage, dynamical diffraction effects, low completeness, missing high resolution reflections, poor signal-to-noise ratio, etc.). For structures with different special positions or high multiplicities, a small coordinate change of one atom on a special position with high multiplicity may change the entire structure, thus it may also significantly vary the R factor due to the number of atoms in the unit cell that are involved. To converge the solution, more weight is usually given to the long-range Coulomb potential energy instead of the contribution from the diffraction data (R factor), otherwise the correct atoms positions are challenging to obtain.

For crystal structures with larger unit cells and numbers of atoms, the number of possible atomic positions in the unit cell (possible configurations) increases exponentially. This causes an exponential increase of computing time for any global optimization method. To speed up structure solution in such cases it is possible to limit the resolution of the data in Endeavour. In contrast to other structure solution methods, this does not necessarily cause a reduced resolution of the structure model since additional information given by the potential energy is used.

Endeavour can (re-)determine the space group of an arrangement of atoms in a unit cell ("configuration") from a triclinic or other crystal system with low space group symmetry once the structure solution is finalized. The correctness of the space group is confirmed by geometrical analysis of the unit cell contents (arrangement of atoms) resulting from the structure solution calculation. In a first step, the symmetry elements in this atomic arrangement are determined, and, in a second step, the space group is tracked down by comparing the resulting set of symmetry elements to all possible space groups. For this, it uses the symmetry finder as it is implemented in the program Kplot developed by R. Hundt et al. [84,85].

Although Endeavour has first been developed for the solution of crystal structures from X-ray or neutron diffraction data, it can also be applied to ED data. Endeavour transforms all diffraction data into a powder diffraction-like format for display purposes, although it internally uses 3D diffraction data for structure solution. The adjustment of Endeavour for electron diffraction is firstly reported here as a tool to be systematically applied to 3D ED data. For this, Endeavour uses atomic form factors calculated from X-ray scattering factors by applying the Mott-Bethe formula [86]. The effect of precession is modelled in Endeavour using a formula provided by P. Moeck [87]. Dynamical effects are not considered in the calculation of the reflection intensities.

It should be noted that the combined cost function approach implemented in Endeavour works best if the potential energy includes long-range interactions like the Coulomb forces between differently charged ions. This eliminates (or at least reduces) the danger of getting trapped in some "unphysical" minimum where, for instance, a lot of cations are placed directly in contact to each other, which is rather unlikely to be observed in nature. This "long-range interaction" effect is of course much less prominent in intermetallic or organic molecule compounds. With metals though, there is another effect that facilitates the solution of crystal structures using the Endeavour approach: the dense packing of atoms (especially at increased pressures) leaves little room for alternatives, thus it would be possible to obtain good quality results when solving crystal structures of intermetallic compounds. However, this situation becomes more complicated for organic molecule structures because the covalent bonds between neutral atoms cause the structures to be less dense packed while at the same time miss the long-range directing interactions of Coulomb forces.

Endeavour can also be used to complete a partial structure model, by fixing the positions of known atoms in the crystal structure and have the software search for a limited number of unknown atom positions. This feature was not utilised in the research presented here and it will be explored in future work.

3. Results and Discussion

The obtained structure models for each of the four materials under study are hereunder presented according to the outputs obtained from the three different crystallographic structure solution methods. Datasets acquired with precession have been processed with ADT3D and PETS2 software packages in order to produce two *hkl* files; one in which the reflection intensity is assigned as the maximum of its rocking curve (ADT3D), and another one where the reflection intensity is extracted as the scale factor of the fitting of a double-peaked function to its rocking curve (PETS2). The former will be referred as max-int file while the latter as fit-int file. Both reflection intensity files are independently used to find structure solutions with the different methods. For comparison purposes, the refined unit cell parameters from X-ray data have been used to run the different programs. Nevertheless, unit cell parameters based on 3D ED data are also provided to show that they are closely following the values of X-ray studies.

3.1. $PrBa_2Cu_3O_7$

The reconstruction of the diffraction space from 61 diffraction patterns acquired with 1° of precession allowed to determine the following unit cell parameters in ADT3D: a = 3.81 Å, b = 3.86 Å, c = 11.61 Å, α = 90.2°, β = 90.0° and γ = 90.2°. These values slightly deviate from the ones reported by Bertrand et al. (a = 3.8605 Å, b = 3.9243 Å, c = 11.6993 Å, α = 90°, β = 90.0° and γ = 90°) [60,66] within the typical 1–5% inaccuracy of unit cell parameters determined using ED data [20]. The main crystallographic sections of the reconstructed 3D ED data are shown in Figure 2, which do not show any systematic extinction following the *Pmmm* space group.

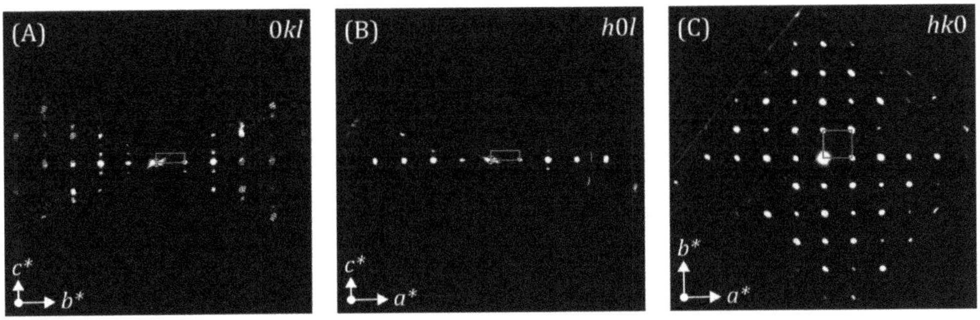

Figure 2. **(A)** 0*kl*, **(B)** *h*0*l* and **(C)** *hk*0 sections of the reconstructed diffraction space of $PrBa_2Cu_3O_7$ from PETS2. The red rectangles represent the projected unit cell along the respective unit cell axes. *a**, *b** and *c** are the reciprocal unit cell axes.

Table 1 shows crystallographic parameters and found atom fractional coordinates of the structure solutions obtained from the different crystallographic methods.

Table 1. Crystallographic experimental parameters of the structure solution of $PrBa_2Cu_3O_7$ and their atomic coordinates obtained by direct methods in SIR2014, charge-flipping in Jana2006 and the global optimization method in Endeavour using the extracted intensities by ADT3D (max-int file) and PETS2 (fit-int file).

	X-ray: Powder [66]	3D ED with PED					
		Direct Methods		Charge-Flipping		Global Optimization	
		ADT3D	PETS2	ADT3D	PETS2	ADT3D	PETS2
Used Resolution (Å)	-	0.8	0.8	0.8	0.8	-	-
Total/Used [1] Refl. (#)	-	425/95	426/96	425/96	426/97	425	426
Completeness (%)	-	40.4	39.8	41.3	40.2	-	-
R_{int} (%) [2]	-	11.4	13.5	11.4	13.5	11.4	13.5
R(F)/R1(obs) (%) [3]	-	10.7	11.0	37.5	56.6	59.35 [5]	60.89 [5]
RMSD [4]	-	0.123	0.079	0.013	0.104	0.130	0.120
	Atomic fractional coordinates (x, y, z)						
Pr1	0.5 0.5 0.5	0.5 0.5 0.5	0.5 0.5 0.5	0.5 0.5 0.5	0.5 0.5 0.5	0.5 0.5 0.5	0.5 0.5 0.5
Ba1	0.5 0.5 0.1804	0.5 0.5 0.165	0.0 0.0 0.159	0.5 0.5 0.177	0.5 0.5 0.164	0.5 0.5 0.185	0.5 0.5 0.185
Cu1	0.0 0.0 0.0	0.0 0.0 0.0	0.0 0.0 0.0	0.0 0.0 0.0	0.0 0.0 0.0	0.0 0.0 0.0	0.0 0.0 0.0
Cu2	0.0 0.0 0.3515	0.0 0.0 0.293	0.0 0.0 0.336	0.0 0.0 0.349	0.0 0.0 0.320	0.0 0.0 0.359	0.0 0.0 0.359
O1	0.0 0.5 0.0	0.0 0.5 0.0	Not found	Not found	Not found	0.0 0.5 0.0	0.0 0.5 0.0
O2	0.0 0.0 0.168	Not found	Not found	Not found	Not found	0.0 0.0 0.181	0.0 0.0 0.178
O3	0 0.5 0.365	Not found	Not found	Not found	Not found	0.0 0.5 0.391	0 0.5 0.390
O4	0.5 0 0.383	0.5 0.0 0.3655	Not found	Not found	Not found	0.5 0.0 0.391	0.5 0 0.389

[1] *Used reflections refer to symmetrically independent reflections up to the used resolution.* [2] *As given by SIR2014 for ADT3D hkl files and as given by PETS2 for PETS2 hkl files.R(F) as given by SIR2014 for direct methods solutions and R1(obs) as given by Jana.* [3] *2006 calculated from the obtained charge-flipping solutions.* [4] *Calculated in SIR2014 only from the found atoms.* [5] *Calculated by Endeavour software after the global optimization of the structure.*

3.1.1. Direct Methods

The use of the max-int file reported all the metal atoms present in the compound and two of the four symmetrically independent oxygens. Three more ghost atoms are given by the program, but they do not make chemical sense and they are more likely related to termination errors as they appear close to special positions. Although both oxygens are very close to their reference positions, one of the Cu atoms is placed 0.7 Å away from the reference one, which would require further refinements including the two missing oxygens to check if this copper is then properly positioned. When the fit-int file is used for comparison, the four metal atoms are correctly spotted with much better accuracy (up to 0.25 Å) than the result from the max-int file, but oxygen positions could not be found, even with several ghost atoms appearing in the solution. Although the *R(F)* value is quite good in both cases (10.7% and 11.0%, respectively), the low completeness of the dataset (around 40%) limits the capability of the algorithm to correctly identify all atom positions.

3.1.2. Charge-Flipping

The use of both extracted intensity files to run charge-flipping reported similar results to the ones obtained from direct methods. For the max-int file, metal atoms are much better positioned (up to 0.04 Å with respect to the reference), although oxygen positions are not located. In the case of the fit-int file, oxygen positions are still not found, and metal atoms are placed with a maximum difference with respect to the reference structure of 0.4 Å. The *R1(obs)* accounts to 37.5% and 56.6%, respectively, which is quite different to the values obtained in SIR2014 but comes as a result of how the *R* factor is calculated in each software package (SIR2014 also includes ghost atoms in the structure model after finishing the execution of direct methods). More interesting is the difference between the *R1(obs)* of the two obtained models in Jana2006. This points out how the higher displacement of the Cu2 atom highly changes the calculated reflection intensities, increases the difference between calculated and observed reflections and thus pushes up the figure-of-merit, providing a hint that some refinement cycles may bring this atomic position to the correct place. Nevertheless, there are no significant differences between these incomplete models and the ones obtained with direct methods, arguing that the low completeness is also a problem for the charge-flipping algorithm.

3.1.3. Global Optimization Method

Extracted reflection intensities from ADT3D and PETS were individually used to run the global optimization method. Unit cell parameters as reported in the literature [60,66], space group *Pmmm*, chemical moiety formula (Pr1Ba2Cu3O7), number of formula units per unit cell (Z = 1), and oxidation states of the atoms (Ba = +2, Pr = +4, O = −2, Cu = +2) were used as input. As reflection intensities may not be too accurate and the structure has many special positions, the potential energy descriptor was given more weight than the *R* factor (0.95 potential energy vs. 0.05 *R* factor). All individual reflections including the Friedel pairs were used instead of any merging of symmetry equivalent reflections.

For both types of extracted intensities, all atomic coordinates (heavy and light atoms) were successfully determined, and correctness of the structures were measured using the root-mean-square deviation (RMSD) with respect to the X-ray reference structure found in the literature (RMSD 0.130 Å for max-int and RMSD 0.120 Å for fit-int). The *R* values for both structures were very similar (*R* factor for max-int: 59.35%, *R* factor for fit-int: 60.89%). The *Pmmm* space group was also correctly confirmed after the determination of the crystal structure. Maximum position shifts observed for one of the O3 atom was 0.3 Å for both structures (Figure 3). It is also worth to point out that the correct atomic positions were also obtained using equal contribution of potential energy vs. *R* factor for both types of extracted intensities, but for better structural accuracy (low RMSD) the higher weight of the potential energy was used during the final global optimization process.

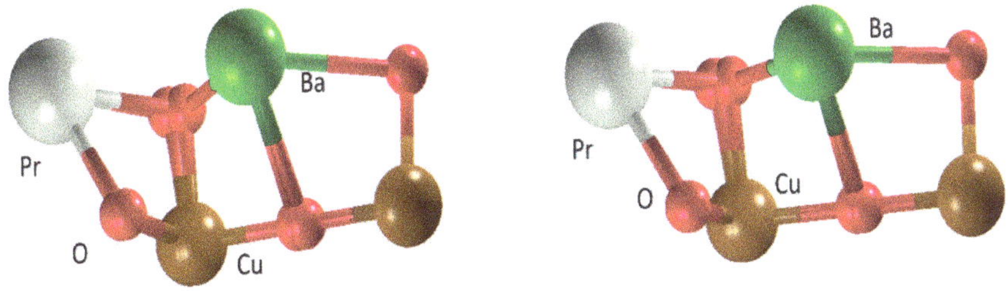

Figure 3. Structural overlays of the X-ray reported structure and the structure obtained by global optimization for the $PrBa_2Cu_3O_7$ compound by using the two differently extracted reflection intensities (**left**: structure from max-int *hkl* file, **right**: structure from fit-int *hkl* file).

3.2. $BaCuO_2$

The reconstruction of the diffraction space from 61 diffraction patterns acquired without the use of precession allowed the determination of the following unit cell parameters in ADT3D: a = 18.20 Å, b = 18.13 Å, c = 18.26 Å, $\alpha = 90.2°$, $\beta = 90.1°$ and $\gamma = 89.3°$. The values are slightly different from the ones reported by Kipka et al. [67] (a = b = c = 18.2700 Å, $\alpha = \beta = \gamma = 90°$) and they are inside the typical 1–5% inaccuracy of unit cell parameters determined using ED data [20]. The 0*kl* and *hhl* crystallographic sections of the reconstructed 3D ED data are shown in Figure 4. The 0*kl* section indicates the reflection condition $h + k = 2n$, while in the *hhl* section the $l = 2n$ is fulfilled, thus matching to the systematic extinctions of the expected $Im\bar{3}m$ space group. It was not possible to collect another dataset with precession from the same sample as beam damage was observed. As there is no data with beam precession, only the max-intensity reflection file was used for structure analysis.

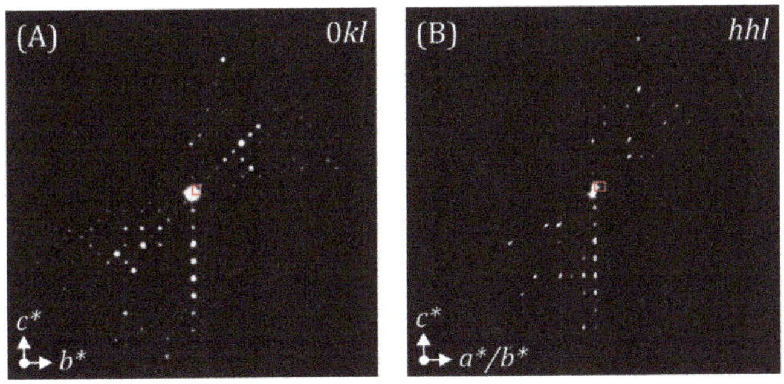

Figure 4. (**A**) 0*kl* and (**B**) *hhl* sections of the reconstructed diffraction space of $BaCu_2$ from PETS2. The red squares represent the projected unit cell along the respective unit cell axes. a^*, b^* and c^* are the reciprocal unit cell axes.

Table 2 shows crystallographic parameters and found atom fractional coordinates of the structure solutions obtained from the different crystallographic methods.

3.2.1. Direct Methods

One of the challenges to obtain structure models from materials like $BaCuO_2$ is to find the positions of partially occupied atoms. In this case, one of the Cu atoms is half-occupied and one of the oxygen atoms is quarter-occupied. Also note that, even if the crystalline structure is highly symmetric, which results in a completeness of 100%, the crystal contains 14 symmetry-independent atoms.

Table 2. Crystallographic experimental parameters of the structure solution of $BaCuO_2$ and their atomic coordinates obtained by direct methods in SIR2014, charge-flipping in Jana2006 and the global optimization method in Endeavour using the extracted intensities by ADT3D (max-int file).

	X-ray: Single-Crystal [67]	3D ED without PED		
		Direct Methods	**Charge-Flipping**	**Global Optimization**
Used Resolution (Å)	-	0.87	0.87	-
Total/Used [1] Refl. (#)	-	14719/514	14719/514	14719
Completeness (%)	-	99.4	100	-
R_{int} (%) [2]	-	60.2		60.2
R(F)/R1(obs) (%) [3]	-	15.6	116.1	112.50 [5]
RMSD [4]	-	0.069	0.382	0.698
		Atomic fractional coordinates (x, y, z, a)		
Ba1	0.0 0.151 0.31	0.0 0.1455 0.3051	0.0 0.2916 0.1831	0 0.1396 0.3283
Ba2	0.0 0.364 0.364	0.0 0.3646 0.3646	0.0 0.3475 0.3475	0.137 0.137 0.5
Ba3	0.177 0.177 0.177	0.1785 0.1785 0.1785	0.1335 0.1335 0.1335	0.1710 0.1710 0.1710
Ba4	0.0 0.0 0.0	Not found	Not found	0.0 0.0 0.0
Cu1	0.25 0.15 0.35	0.25 0.1519 0.3481	0.2659 0.1014 0.3827	0.25 0.1445 0.3555
Cu2	0.0 0.125 0.125	0.0 0.1177 0.1177	Not found	0.0 0.1281 0.1281
Cu3	0.206 0.0 0.0	0.2259 0.0 0.0	0.1502 0.0 0.0	-
Cu4	0.43 0.0 0.0 (0.5)	0.4302 0.0 0.0 (1)	0.3701 0.0 0.0 (1)	-
O1	0.072 0.072 0.186	Not found	0.0911 0.0911 0.2380	-
O2	0.144 0.144 0.343	0.1488 0.1488 0.3151	Not found	0.1497 0.1497 0.3283
O3	0.267 0.267 0.085	0.2637 0.2637 0.0897	Not found	-
O4	0.25 0.0 0.5	0.25 0.0 0.5	Not found	0.25 0.0 0.5
O5	0.338 0.0 0.0	Not found	Not found	-
O6	0.0 0.112 0.44 (0.25)	Not found	Not found	-

[1] *Used reflections refer to symmetrically independent reflections up to the used resolution.* [2] *As given by SIR2014 for ADT3D hkl files and as given by PETS2 for PETS2 hkl files.* [3] *R(F) as given by SIR2014 for direct methods solutions and R1(obs) as given by Jana2006 calculated from the obtained charge-flipping solutions.* [4] *Calculated in SIR2014 only from the found atoms.* [5] *Calculated by Endeavour software after the global optimization of the structure.*

The retrieval of a structure solution as complete as possible with respect to the reference structure was not an easy task in this case. After some trials of slight changes in composition and resolution used for direct methods, a partial structure model could be obtained. The correct composition of 90 Ba, 90 Cu and 180 O that takes into account the partially occupied elements and a resolution of 0.87 Å were used. In this way, three of the six O positions, three of the four Ba positions and all Cu positions were found in a structure solution given by SIR2014 that contained 15 assigned atoms, five of which were ghost ones.

The *R(F)* for the obtained structure model (including ghost atoms) was 15.6%, but a closer look to how this solution resembles to the reference structure shows a high position deviation for some atoms. The maximum deviation for the found oxygens is about 0.5 Å, 0.4 Å for copper atoms and 0.1 Å for barium ones, a difference that can be confirmed by the high RMSD. The accuracy of the derived model could be improved by some refinement cycles. Furthermore, the high R_{int} value reported by SIR2014 demonstrates how the reflection intensities poorly follow the symmetry relations of the $m\bar{3}m$ Laue class. Low dynamic range of the detector with poor signal-to-noise ratio, beam damage and the possible dynamical effects due to beam precession absence might be associated with the poor data quality, which ultimately gets reflected in the obtained structure model.

3.2.2. Charge-Flipping

After several attempts with Superflip, it was found that a resolution of 0.87 Å and a fixed atomic displacement parameter (ADP) of 2 $Å^2$ allowed to obtain a partial structure solution. In this case, most of the heavy elements were identified but only one of the six oxygens could be identified. However, the majority of the atom positions were deviated between 1.0 Å and 1.4 Å with respect to the reference model. In some cases, atoms had a high displacement close to partially occupied or special positions. For instance, Cu1 should be sitting in a special position, but Superflip assigned its position to a peak near the special one, resulting in a position shift of 1.1 Å. In another case, Cu4 should be a partially occupied atom, but the program assigns a fully occupied atom between its correct position and the close-by and fully occupied oxygen (O5), which leads to a final mismatch of 1.1 Å.

Owing to this highly inaccurate structure model, a final *R1(obs)* of 116.1% is obtained as well as a very high RMSD, which indicates how wrong is the structure with respect to the reference one and shows that, in this example, the structure solution from direct methods is better.

3.2.3. Global Optimization Method

The unit cell parameters as reported in the literature [67], space group $I\,m\bar{3}m$, chemical moiety formula (Ba1 Cu1 O2), number of formula units per unit cell (Z = 90), atom oxidation states (Ba = +2, Cu = +2, O = −2) and extracted reflection intensities were used as input for global optimization. During the structure solution process, no further merging of symmetry equivalent reflections including the Friedel pairs was performed. Due to low quality data and a space group with several special positions and high multiplicity, a higher weight was given to the potential energy descriptor rather than the *R* factor (0.95 potential energy vs. 0.05 *R* factor).

Another challenge for $BaCuO_2$ structure was that there are atomic positions with partial occupation. This feature is currently not yet available for structure solution calculations in Endeavour. For the moment, one has to assume fully occupied positions. Unfortunately, it normally only works if the change of electrostatic potential at a certain position is not overly large. In the case of $BaCuO_2$, there is a half-occupied Cu placed on the special position 12e, so increasing the site occupation factor of this Cu from 0.5 to 1.0 will significantly change the scattering density and hence the reflection intensities. Furthermore, structure solutions with larger unit cell dimensions by global optimization methods take more time to complete. The procedure was sped up by using a restricted dataset that only included reflections up to a 2θ value of 1°, which corresponds to a minimum d-spacing of about 1.44 Å (calculated for a wavelength of 0.0251 Å, the equivalent of electrons accelerated at 200 kV). During the structure solution process some of the ghost atom positions were removed and one of the Cu atom positions were renamed as O atom positions. By increasing the influence of the potential energy and the use of a restricted number of reflection intensities, it was possible to determine all the Ba atomic positions, two Cu atoms and two O atom positions successfully (RMSD 0.698 Å). This final solution took approximately 17 h on a Windows PC with Pentium i3 processors at 3.9 GHz.

3.3. $LiAl_{0.8}Fe_{0.2}(SiO_3)_2$

The reconstruction of the diffraction space from 58 diffraction patterns acquired with 1° of precession allowed the determination of the following unit cell parameters in ADT3D: a = 9.42 Å, b = 8.49 Å, c = 5.28 Å, α = 89.9°, β = 109.9° and γ = 90.0°. Such values are slightly different from the ones reported by Iezzi et al. [65] (a = 9.490 Å, b = 8.426 Å, c = 5.232 Å and β = 110.14°) and they are inside the typical 1–5% error of electron diffraction [20]. The main crystallographic sections of the reconstructed 3D ED data are shown in Figure 5. The *0kl* crystallographic section exhibits the reflection condition $k = 2n$, while the *h0l* fulfils the following condition $h, l = 2n$. Unfortunately, the *hk*0 section does not have enough information to properly assert that systematic extinctions appear when $h + k \neq 2n$, which would confirm the expected $C\,2/c$ space group. When dealing with unknown structures, the large missing cone of low angular range 3D ED datasets may cause difficulties in the symmetry determination of the crystal under study.

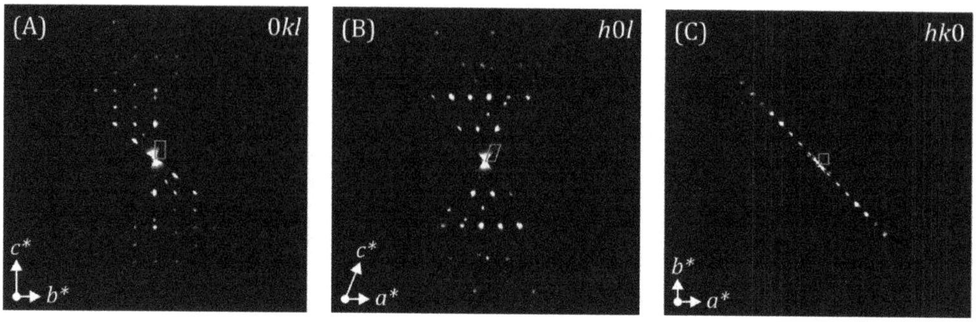

Figure 5. (**A**) 0*kl*, (**B**) *h*0*l* and (**C**) *hk*0 sections of the reconstructed diffraction space of $LiAl_{0.8}Fe_{0.2}(SiO_3)_2$ from PETS2. The red markers represent the projected unit cell along the respective unit cell axes. *a**, *b** and *c** are the reciprocal unit cell axes.

Table 3 shows crystallographic parameters and found atom fractional coordinates of the structure solutions obtained from the different crystallographic methods.

Table 3. Crystallographic experimental parameters of the structure solution of $LiAl_{0.8}Fe_{0.2}(SiO_3)_2$ and their atomic coordinates obtained by direct methods in SIR2014, charge-flipping in Jana2006 and the global optimization method in Endeavour using the extracted intensities by ADT3D (max-int file) and PETS2 (fit-int file).

	X-ray: Single-Crystal [65]	3D ED with PED					
		Direct Methods		Charge-Flipping		Global Optimization	
		ADT3D	PETS2	ADT3D	PETS2	ADT3D	PETS2
Used Resolution (Å)	-	0.7	0.7	0.7	0.7	-	-
Total/Used [1] Refl. (#)	-	7007/312	1670/322	7007/313	1670/323	7007	1670
Completeness (%)	-	52.7	54.4	53.3	54.8	-	-
R_{int} (%) [2]	-	10.0	19.3	10.0	19.3	10.0	19.3
R(F)/R1(obs) (%) [3]	-	16.5	15.3	71.8	65.8	94.27 [5]	76.32 [5]
RMSD [4]	-	0.026	0.015	0.063	0.020	0.056	0.105
Atomic fractional coordinates (x, y, z)							
Li1	0.0 0.2746 0.25	Not found	Not found	Not found	Not found	0.0 0.2734 0.25	0.0 0.2609 0.25
Al1/Fe1	0.0 0.9067 0.25	0.0 0.0973 0.75	0.0 0.0981 0.75	0.0 0.0989 0.75	0.0 0.0980 0.75	0.500 0.404 0.25	0.500 0.408 0.25
Si1	0.2941 0.0935 0.2559	0.2926 0.0904 0.2509	0.2948 0.0913 0.2544	0.2922 0.0983 0.2491	0.2901 0.0937 0.2528	0.2956 0.0905 0.2572	0.3003 0.0888 0.2648
O1	0.1096 0.0825 0.1404	0.1012 0.0851 0.1385	0.1064 0.0809 0.1414	0.1157 0.0726 0.1488	0.1091 0.0803 0.1412	0.1172 0.0896 0.1483	0.1141 0.0836 0.1499
O2	0.3647 0.2669 0.3004	0.3737 0.2638 0.3040	0.3682 0.2635 0.3032	0.3800 0.2613 0.2826	0.3732 0.2644 0.3040	0.3589 0.2620 0.2890	0.3619 0.2560 0.2933
O3	0.3565 0.9867 0.0585	0.3546 0.9872 0.0549	0.3569 0.9876 0.0566	Not found	0.3561 0.9870 0.0558	0.3537 −0.014 0.0625	0.3532 −0.013 0.0576

[1] *Used reflections refer to symmetrically independent reflections up to the used resolution.* [2] *As given by SIR2014 for ADT3D hkl files and as given by PETS2 for PETS2 hkl files.* [3] *R(F) as given by SIR2014 for direct methods solutions and R1(obs) as given by Jana2006 calculated from the obtained charge-flipping solutions.* [4] *Calculated in SIR2014 only from the found atoms.* [5] *Calculated by Endeavour software after the global optimization of the structure.*

3.3.1. Direct Methods

The elemental composition provided to direct methods is a key factor for a correct structure solution from ED data because it is associated with the scattering factors that have to be used for the crystal structure determination procedure. However, a composition of 0.8Al + 0.2Fe, 1.0Al or 1.0Fe in the octahedron position did not significantly change the outcome of this method, thus a fully occupied iron (Fe) position was used. Furthermore, neither direct methods nor charge-flipping are able to give ab initio partially occupied positions, which leaves this structural characterization step to structure refinement.

By using the max-int file or the fit-int file, the whole structure model could be retrieved except for the lithium positions. In both cases, more than 10 ghost atoms are given by SIR2014, but none of them corresponded to the accurate position of lithium. A maximum atom position difference of 0.09 Å is found for one of the oxygens of the structure model related to the max-int file, while a value of 0.04 Å is found for two of the oxygens and iron for the solution of the fit-int file.

The *R(F)* values are 16.5% and 15.3% for the max-int and fit-int structure solutions, respectively. Since both structure models contain the same amount of ghost atoms, it is clear that the difference between them comes from the better accuracy of the model obtained from the fit-int file, also confirmed by the RMSD (0.015 against 0.026).

3.3.2. Charge-Flipping

Charge-flipping was initially performed with the max-int file and a fixed composition for peak search with a fully occupied iron (Fe) in the octahedron position. All atoms were found except for one of the oxygens and the lithium, which Superflip assigned to other peaks from the calculated electrostatic potential. However, one of the correctly spotted oxygens is deviated about 0.2 Å with respect to the reference position. When the fit-int file is instead used, the obtained structure solution is enhanced. All non-lithium atoms are properly located with a maximum deviation of 0.08 Å (corresponding to one of the oxygens) in comparison to the reference model.

The *R1(obs)* is 71.8% for max-int reflections and 65.8% for fit-int reflections. Such high values are obtained mainly because atoms do not have ADP values that properly fit the electrostatic potential, and they would require further refinement steps to improve it. It is also worth to point out how the RMSD jumps for the max-int solution due to the highly deviated oxygen (0.063 against 0.020).

3.3.3. Global Optimization Method

The unit cell parameters as reported in the literature [63], space group $C\,2/c$, chemical moiety formula (Li1 Al1 Si2 O6), number of formula units per unit cell (Z = 4), atom oxidation states (Li = +1, Al= +3, Si = +4, O = −2) and both extracted reflection intensities were used as input for global optimization. As previous optimization processes, no further merging of symmetry equivalent reflections and Friedel pair reflections was performed. Though the structure has partially occupied Al positions, Endeavour cannot accommodate partial occupancy for the global optimization, so it was assumed that the structure does not contain any Fe atom. Since the ratio of Fe/Al is rather low (1:4) and the electron scattering coefficients of Fe and Al are not highly different, the error introduced by replacing the Fe atoms with Al is limited with respect to the scattering power at the corresponding atomic position. During the initial global optimization, equal weight was given to the Columbic potential energy and the *R* factor, but it turns out that either there are many additional ghost atoms in the structure, or the entire structure solution did not converge to the right model. This could be because of the low quality of the data, thus a new global optimization was performed using higher weight to the potential energy (0.95 potential energy vs. 0.05 *R* factor). With this new weight scheme, structure solution converged, and correct models were obtained from both extracted reflection intensities. Interestingly, Li atom position was also determined, which is always a challenge from direct methods or charge-flipping as Li atoms are mobile in nature with low scattering power. A RMSD of 0.056 Å was obtained for the structure using the max-int reflections, whereas the structure obtained using the fit-int reflections have a RMSD of 0.105 Å (Figure 6). Although the atomic structures are correct, high *R* values (94.27% for max-int reflections and 76.32% for fit-int reflections) were obtained for the final model in the global optimization process.

The latter is potentially related to noise from the high-resolution reflections which has been included during the global optimization process and finally used to calculate the *R* value of the obtained solution from global optimization process. Additionally, *C2/c* symmetry of the solved structure after global optimization process has also been confirmed by the symmetry finder module available in Endeavour.

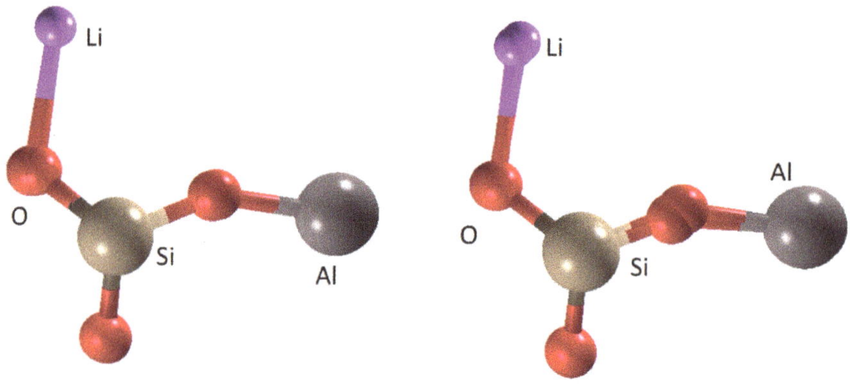

Figure 6. Structural overlays of the X-ray reported structure and the structure obtained by global optimization for the $LiAl_{0.8}Fe_{0.2}(SiO_3)_2$ compound by using the two differently extracted reflection intensities (**left**: structure from the max-int *hkl* file, **right**: structure from fit-int *hkl* file).

3.4. $LiAl(SiO_3)_2$

The reconstruction of the diffraction space from 61 diffraction patterns acquired with 1° of precession allowed the determination of the following unit cell parameters in ADT3D: a = 9.97 Å, b = 8.67 Å, c = 5.478 Å, α = 89.5°, β = 111.3° and γ = 90.3°. Such values do not deviate much from the ones reported by Redhammer et al. [63] (a = 9.474 Å, b = 8.390 Å, c = 5.219 Å and β = 110.07°) and they are inside the typical 1–5% inaccuracy of unit cell determination from ED data [19]. The main crystallographic sections of the reconstructed 3D ED data are shown in Figure 7. The 0*kl* section shows that reflections appear when $k = 2n$, the $h, l = 2n$ reflection condition is fulfilled for the *h*0*l* section, and the *hk*0 section exhibits systematic extinctions that belong to the reflection condition $h + k = 2n$, therefore confirming the expected *C*2/*c* space group. In comparison to the dataset of $LiAl_{0.8}Fe_{0.2}(SiO_3)_2$, the orientation of the crystal with respect to the geometry of the experimental setup was in such a way that enough information was collected from the main reciprocal sections, which allows here to properly determine the symmetry of the crystal.

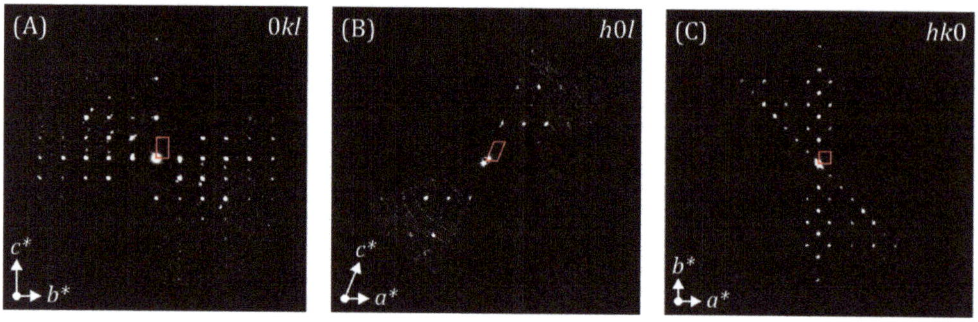

Figure 7. (**A**) 0*kl*, (**B**) *h*0*l* and (**C**) *hk*0 sections of the reconstructed diffraction space of $LiAl(SiO_3)_2$ from PETS2. The red markers represent the projected unit cell along the respective unit cell axes. *a**, *b** and *c** are the reciprocal unit cell axes.

Table 4 shows crystallographic parameters and found atom fractional coordinates of the structure solutions obtained from the different crystallographic methods.

Table 4. Crystallographic experimental parameters of the structure solution of $LiAl(SiO_3)_2$ and their atomic coordinates obtained by direct methods in SIR2014, charge-flipping in Jana2006 and the global optimization method in Endeavour using the extracted intensities by ADT3D (max-int file) and PETS2 (fit-int file).

	X-ray: Single-Crystal [63]	3D ED with PED					
		Direct Methods		Charge-Flipping		Global Optimization	
		ADT3D	PETS2	ADT3D	PETS2	ADT3D	PETS
Used Resolution (Å)	-	0.7	0.7	0.7	0.7	-	-
Total/Used [1] Refl. (#)	-	1597/256	749/261	1597/257	749/262	1597	749
Completeness (%)	-	44.8	45.7	46.7	47.4	-	-
R_{int} (%) [2]	-	11.8	13.1	11.8	13.1	11.8	13.1
R(F)/R1(obs) (%) [3]	-	12.8	9.3	38.0	48.9	46.59 [5]	44.82 [5]
RMSD [4]	-	0.017	0.025	0.024	0.017	0.035	0.027
Atomic fractional coordinates (x, y, z)							
Li1	0.0 0.2746 0.25	0.0 0.2688 0.25	0.0 0.2791 0.25	Not found	Not found	0.0 0.2615 0.25	0.0 0.2725 0.25
Al1	0.0 0.9067 0.25	0.0 0.9029 0.25	0.0 0.9066 0.25	0.0 0.9065 0.25	0.0 0.9044 0.25	0.500 0.408 0.246	0.500 0.407 0.246
Si1	0.2941 0.0935 0.2559	0.2980 0.0951 0.2555	0.2974 0.0945 0.2517	0.2892 0.0951 0.2630	0.2917 0.0941 0.2565	0.2899 0.0942 0.2690	0.2941 0.0944 0.2720
O1	0.1096 0.0825 0.1404	0.1040 0.0856 0.1378	0.1096 0.0830 0.1400	0.1019 0.0828 0.1420	0.1049 0.0844 0.1430	0.1137 0.0840 0.1310	0.1129 0.0846 0.1250
O2	0.3647 0.2669 0.3004	0.3647 0.2691 0.2972	0.3565 0.2658 0.3073	0.3628 0.2666 0.2999	0.3646 0.2663 0.3014	0.3586 0.2665 0.2980	0.3611 0.2652 0.2960
O3	0.3565 0.9867 0.0585	0.3555 0.9912 0.0598	0.3628 0.9872 0.0469	0.3601 0.9920 0.0579	0.3618 0.9914 0.0598	0.3492 −0.016 0.0697	0.3543 −0.022 0.0621

[1] Used reflections refer to symmetrically independent reflections up to the used resolution. [2] As given by SIR2014 for ADT3D hkl files and as given by PETS2 for PETS2 hkl files. [3] R(F) as given by SIR2014 for direct methods solutions and R1(obs) as given by Jana2006 calculated from the obtained charge-flipping solutions. [4] Calculated in SIR2014 only from the found atoms. [5] Calculated by Endeavour software after the global optimization of the structure.

3.4.1. Direct Methods

The max-int file was first used to retrieve the structure model. Although twelve ghost atoms were given by SIR2014, silicon, aluminum and oxygen atoms were correctly found with electrostatic potential peaks significantly higher than the remaining atomic positions, and a maximum deviation with respect to the reference structure model of 0.06 Å, which correspond to one of the oxygens. The lithium was spotted to be one of the atoms with much lower potential values and a position difference of 0.05 Å in comparison to the reference one.

The result of the fit-int file is quite similar to the one of max-int file. In this case, the six highest peaks obtained by SIR2014 correspond to the six atoms of the structure model. The software package correctly assigned the element for each atom. A maximum position difference of 0.1 Å is detected for two of the oxygens, while lithium is placed very closed to the reference one (0.04 Å).

The *R(F)* values obtained from both structures (including ghost atoms) are 12.8% and 9.3%, respectively. Here it is interesting to note that the *R(F)* is better for the fit-int file even if the structure model is slightly worse with respect to the reference one. In comparison to the result of $PrBa_2Cu_3O_7$, the low completeness does not seem to be a limiting factor for finding a correct structure solution, and the higher quality of the reflection intensities helps to elucidate a proper crystalline model.

3.4.2. Charge-Flipping

Initially, the max-int file was used to retrieve structure models. After changing the default ADP to 2 $Å^2$ (B_{iso}) and selecting a fixed composition for peak search, a structure solution with all atoms except lithium could be found with a maximum deviation of 0.08 Å (corresponding to one of the oxygens) with respect to the reference positions. SUPERFLIP assigned the lithium to other maxima of the calculated electrostatic potential but far away from the correct one. This has been also the case for SIR2014 when dealing with the max-int file, but the availability of other ghost atoms in the given structure solutions allowed the possibility to properly locate it.

The extracted results from the fit-int file were not different to the ones obtained with the max-int file, although the default value for the ADP (0 $Å^2$) was kept in this case. A maximum deviation with respect to the reference structure was found to be 0.06 Å for one of the oxygens.

The final *R1(obs)* for these structure models are 38.0% and 48.9%, respectively, and, similar to the results from direct methods, the low completeness doesn't prohibit to properly obtain accurate structure models. One of the differences is that the parameters of SUPERFLIP (method to assign atom positions, ADP value and resolution) had to be tweaked more than in SIR2014 so that the program could correctly interpret and assign the peaks of the electrostatic potential.

3.4.3. Global Optimization Method

The same global optimization process has been followed for $LiAl(SiO_3)_2$ as described previously for $LiAl_{0.8}Fe_{0.2}(SiO_3)_2$. The only difference between these compounds is that for the current structure, $LiAl(SiO_3)_2$, all atom positions are fully occupied. It was also necessary here to set higher weight for the Coulomb potential energy (0.95 potential energy vs. 0.05 *R* factor), otherwise the structure solution did not converge to the right model. In this way, it was also possible to successfully determine all atom positions, including the light Li atom positions for both type of extracted intensity (RMSD of 0.035 Å for the max-int intensity file and RMSD 0.027 Å for the fit-int intensity file) (Figure 8). Interestingly, *R* values for the final structures were much lower in comparison to $LiAl_{0.8}Fe_{0.2}(SiO_3)_2$ (46.59% for max-int *hkl* file and 44.82% for fit-int *hkl* file). The possible difference in *R* values may come due to the better reflection intensities quality obtained using a superior camera, which also allowed a data acquisition procedure in low dose conditions to minimize beam damage. Furthermore, the use of STEM mode for crystal search and data acquisition also facilitates and reduces redundant illumination of targeted crystals, a feature that is essential in order to get suitable data for crystal structure determination of beam sensitive materials such as Li-containing compounds [17,18].

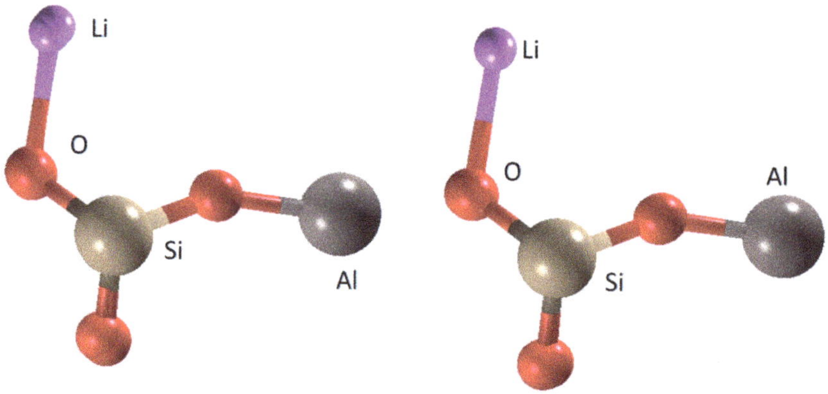

Figure 8. Structural overlays of the X-ray reported structure and the models obtained by global optimization of $LiAl(SiO_3)_2$ by using the differently extracted reflection intensities (**left**: structure from max-int *hkl* file, **right**: structure from fit-int *hkl* file).

4. Conclusions

Four inorganic materials have been studied by means of 3D ED data with different degrees of completeness and quality for crystal structure determination. Each compound was evaluated by three different structure solution algorithms (direct methods, charge-flipping and global optimization), and two different intensity extraction methods were also compared for three of the datasets. Analysis of the structure solution success was given by comparing the models from ED data with the corresponding phases obtained from X-ray diffraction found in the literature.

The global optimization method as implemented in Endeavour successfully found all atom positions for $LiAl_{0.8}Fe_{0.2}(SiO_3)_2$, $LiAl(SiO_3)_2$ and $PrBa_2Cu_3O_7$, including the lithium ones. However, this method is not designed to handle crystal structures with partially occupied atom positions, like $BaCuO_2$, which may lead to spurious atom positions in the structure solution, thus challenges in terms of finding an initial model. Furthermore, structures that have high symmetry and unit cells with at least one long axis become computationally expensive and, at the end, considerably slows down the search for the global optimum.

The results presented here also show that a robust and successful ab initio structure solution by direct methods is possible for beam sensitive Li-bearing compounds even from datasets with low completeness (~45%), provided that the data is acquired in low dose conditions with a higher signal-to-noise electron detector. It has also been observed that in these conditions crystal structure determination from direct methods is slightly more robust than charge-flipping, which may be expected given the statistical foundation of the direct methods algorithm. Another way to improve the chance of a successful structure solution would be to merge different datasets of the same material to push up the completeness, but this was out of the scope in this work and it has been already reported [88,89].

The more challenging dataset analysed here was the case of $BaCuO_2$, where even the heavy atom positions were not successfully determined with any of the tested methods. This could be explained by its occupational disorder, the 3D ED collection without the use of precession and the poor data quality of the detector. One way to increase the chances to retrieve a successful model would be to acquire such data with the use of precession, since it has been demonstrated that the integration of reflections by means of precession helps to minimize dynamical effects [90] thus reflections could follow better the kinematical theory of diffraction from which direct methods, charge-flipping and global optimization are based on. However, these algorithms are not designed to give ab initio partially occupied atoms and it may still require to refine the structure solutions in order to properly describe the material.

The global optimization method can be considered as an alternative crystallographic structure solution method for 3D ED data by adjusting the weighting of the potential energy with respect to the R factor. It is important to keep in mind that the primary goal of this method should not be the chase for an extremely low R value. Instead, one is looking for a chemically and physically reasonable structure model for which the calculated R value is the lowest of all candidate structures (i.e., solutions obtained from multiple Endeavour runs with different seed values). The structure analyses carried out in this work showed that even when the R value of a reasonable structure model is around 40–50%, the R values of competing solutions are typically still larger (e.g., 70–80% or even more), thus the relatively best R value is what one has to look for. However, an R value of 40–50% can be reached by a large variety of arbitrary atomic arrangements, but most of them are not reasonable from a physical/chemical point of view. For this reason, the inclusion of the potential energy in the cost function helps to eliminate this large majority of unreasonable structure models. At this point one could even think of completely omitting the diffraction data and optimizing just the potential energy of the system, i.e., to predict the crystal structure. While this is certainly possible in Endeavour, it turns out that in many cases the resulting structure models are still somewhat wrong. They may already be similar to the correct crystal structure, but there are typically still some details missing. The main reason for this is the extremely simple

2-body-function used for the calculation of the potential energy (Coulomb interactions and minimum interatomic distances, as shown in the Supplementary Materials) [91–93]. The addition of the *R* value linked to the experimental diffraction data in the overall cost function helps to drive the model into the right direction, even if reflection intensities deviate from their theoretically expected values. In other words, the use of experimental data enables the bridge from crystal structure prediction to crystal structure determination.

Global optimization methods in general are computationally expensive techniques, therefore their use should be recommended for the edge cases where direct methods or charge-flipping fail to yield a structure. The presented results show how data acquired with older TEMs and detectors can be utilised to successfully obtain structures of inorganic compounds via the application of global optimisation methods. Ultimately, this increases the accessibility of instruments that can be used to collect 3D ED data for crystal structure determination of nanocrystals.

Supplementary Materials: The following supporting information can be downloaded at: https://www.mdpi.com/article/10.3390/sym14112355/s1.

Author Contributions: P.P.D., S.P.-R., A.S.G., A.S., F.K. and H.P. participated in data collection and processing. Conceptualization, S.N., G.I. and I.M.; methodology, I.M., G.I. and M.C.; validation, M.C.; resources, S.N. and I.M.; data curation, G.I.; writing—review and editing, S.N.; funding acquisition, S.N. All authors contributed to the writing and editing of the manuscript. All authors have read and agreed to the published version of the manuscript.

Funding: NanoMEGAS SPRL which acknowledges funding support for this research work from the European Union Seventh Framework Programme under grant agreement 312483–ESTEEM2 (Integrated Infrastructure Initiative-I3). G.I. was supported by the "Fondi Ateneo of the University G. D'Annunzio" and PRIN (2017J277S9_003) project "Time Scales of Solidification in Magmas: Application to Volcanic Eruptions, Silicate Melts, Glasses, Glass-Ceramics." I.M. acknowledges this research was supported by the Hellenic Foundation for Research and Innovation (HFRI) under the 'First Call for HFRI Research Projects to Support Faculty Members and Researchers and the Procurement of High-Cost Research Equipment Grant' (project No. 3051).

Institutional Review Board Statement: Not applicable.

Informed Consent Statement: Not applicable.

Data Availability Statement: Not applicable.

Acknowledgments: NanoMEGAS SPRL would like to thank the laboratory of Electron Microscopy & Microanalysis, School of Natural Sciences, University of Patras, Greece, for support during this study.

Conflicts of Interest: The authors declare no conflict of interest.

References

1. Cowley, J.M. *Electron Diffraction Techniques*; Oxford University Press: Oxford, UK, 1992; Volumes 1 and 2.
2. Dorset, D.L. *Structural Electron Crystallography*; Plenum Press: New York, NY, USA, 1995.
3. Dorset, D.L.; Hauptman, H.A. Direct phase determination for quasi-kinematical electron diffraction intensity data from organic microcrystals. *Ultramicroscopy* **1976**, *1*, 195–201. [CrossRef]
4. Dorset, D.L. Electron crystallography. *Acta Crystallogr. Sect. B Struct. Sci.* **1996**, *52*, 753–769. [CrossRef]
5. Zhou, X.; Hovmöller, S. Electron crystallography: Imaging and single-crystal diffraction from powders. *Acta Crystallogr. Sect. A* **2008**, *64*, 149–160. [CrossRef]
6. Sinkler, W.; Own, C.S.; Marks, L.D. Application of a 2-beam model for improving the structure factors from precession electron diffraction intensities. *Ultramicroscopy* **2007**, *107*, 543–550. [CrossRef] [PubMed]
7. Ciston, J.; Deng, B.; Marks, L.D.; Own, C.S.; Sinkler, W. A quantitative analysis of the cone-angle dependence in precession electron diffraction. *Ultramicroscopy* **2008**, *108*, 514–522. [CrossRef] [PubMed]
8. White, T.A.; Eggeman, A.S.; Midgley, P.A. Is precession electron diffraction kinematical? Part I:: "Phase-scrambling" multislice simulations. *Ultramicroscopy* **2010**, *110*, 763–770. [CrossRef]
9. Palatinus, L.; Jacob, D.; Cuvillier, P.; Klementová, M.; Sinkler, W.; Marks, L.D. Structure refinement from precession electron diffraction data. *Acta Crystallogr. Sect. A* **2013**, *69*, 171–188. [CrossRef]

10. Vincent, R.; Midgley, P.A. Double conical beam-rocking system for measurement of integrated electron diffraction intensities. *Ultramicroscopy* **1994**, *53*, 271–282. [CrossRef]
11. Own, C.S. System Design and Verification of the Precession Electron Diffraction Technique. Ph.D. Thesis, Northwestern University, Evanston, IL, USA, 2005.
12. Nicolopoulos, S.; Weirich, T.E. ELCRYST 2005 proceedings of the electron crystallography school 2005: New frontiers in electron crystallography. *Ultramicroscopy* **2007**, *107*, 431–558.
13. Available online: www.nanomegas.com (accessed on 10 June 2022).
14. Portillo, J.; Rauch, E.F.; Nicolopoulos, S.; Gemmi, M.; Bultreys, D. Precession Electron Diffraction Assisted Orientation Mapping in the Transmission Electron Microscope. *Mater. Sci. Forum* **2010**, *644*, 1–7. [CrossRef]
15. Darbal, A.D.; Narayan, R.D.; Vartuli, C.; Lian, G.; Graham, R.; Shaapur, F.; Nicolopoulos, S.; Weiss, J.K. Automated High Precision Strain Measurement Using Nanobeam Diffraction Coupled with Precession. *Microsc. Microanal.* **2013**, *19*, 702–703. [CrossRef]
16. Hoque, M.M.; Vergara, S.; Das, P.P.; Ugarte, D.; Santiago, U.; Kumara, C.; Whetten, R.L.; Dass, A.; Ponce, A. Structural Analysis of Ligand-Protected Smaller Metallic Nanocrystals by Atomic Pair Distribution Function under Precession Electron Diffraction. *J. Phys. Chem. C* **2019**, *123*, 19894–19902. [CrossRef]
17. Kolb, U.; Gorelik, T.; Kübel, C.; Otten, M.T.; Hubert, D. Towards automated diffraction tomography: Part I—Data acquisition. *Ultramicroscopy* **2007**, *107*, 507–513. [CrossRef]
18. Kolb, U.; Gorelik, T.; Otten, M.T. Towards automated diffraction tomography. Part II—Cell parameter determination. *Ultramicroscopy* **2008**, *108*, 763–772. [CrossRef] [PubMed]
19. Mugnaioli, E.; Gorelik, T.; Kolb, U. "Ab initio" structure solution from electron diffraction data obtained by a combination of automated diffraction tomography and precession technique. *Ultramicroscopy* **2009**, *109*, 758–765. [CrossRef] [PubMed]
20. Kolb, U.; Mugnaioli, E.; Gorelik, T.E. Automated electron diffraction tomography—A new tool for nano crystal structure analysis. *Cryst. Res. Technol.* **2011**, *46*, 542–554. [CrossRef]
21. Kolb, U.; Gorelik, T.; Mugnaioli, E. Automated diffraction tomography combined with electron precession: A new tool for ab initio nanostructure analysis. *MRS Online Proc. Libr.* **2009**, *1184*, 38–50. [CrossRef]
22. Zhang, D.; Oleynikov, P.; Hovmoller, S.; Zou, X. Collecting 3D electron diffraction data by the rotation method. *Z. Krist. Cryst. Mater.* **2010**, *225*, 94–102. [CrossRef]
23. Wan, W.; Sun, J.; Su, J.; Hovmöller, S.; Zou, X. Three-dimensional rotation electron diffraction: Software RED for automated data collection and data processing. *Appl. Crystallogr.* **2013**, *46*, 1863–1873. [CrossRef]
24. Palatinus, L.; Brázda, P.; Jelínek, M.; Hrdá, J.; Steciuk, G.; Klementová, M. Specifics of the data processing of precession electron diffraction tomography data and their implementation in the program PETS2.0. *Acta Crystallogr. Sect. B* **2019**, *75*, 512–522. [CrossRef]
25. Altomare, A.; Burla, M.C.; Camalli, M.; Cascarano, G.L.; Giacovazzo, C.; Guagliardi, A.; Moliterni, A.G.G.; Polidori, G.; Spagna, R. SIR97: A new tool for crystal structure determination and refinement. *J. Appl. Crystallogr.* **1999**, *32*, 115–119. [CrossRef]
26. Altomare, A.; Corriero, N.; Cuocci, C.; Moliterni, A.; Rizzi, R. The hybrid big bang-big crunch method for solving crystal structure from powder diffraction data. *J. Appl. Crystallogr.* **2013**, *46*, 779–787. [CrossRef]
27. Weirich, T.E.; Zou, X.D.; Ramlau, R.; Simon, A.; Cascarano, G.L.; Giacovazzo, C.; Hovmöller, S. Structures of nanometre-size crystals determined from selected-area electron diffraction data. *Acta Crystallogr. Sect. A* **2000**, *56*, 29–35. [CrossRef]
28. Palatinus, L.; Chapuis, G. SUPERFLIP—A computer program for the solution of crystal structures by charge flipping in arbitrary dimensions. *J. Appl. Crystallogr.* **2007**, *40*, 786–790. [CrossRef]
29. Palatinus, L. The charge-flipping algorithm in crystallography. *Acta Crystallogr. Sect. B Struct. Sci. Cryst. Eng. Mater.* **2013**, *69*, 1–16. [CrossRef]
30. Nannenga, B.L.; Shi, D.; Leslie, A.G.W.; Gonen, T. High-resolution structure determination by continuous-rotation data collection in MicroED. *Nat. Methods* **2014**, *11*, 927–930. [CrossRef]
31. Nederlof, I.; van Genderen, E.; Li, Y.-W.; Abrahams, J.P. A Medipix quantum area detector allows rotation electron diffraction data collection from submicrometre three-dimensional protein crystals. *Acta Crystallogr. Sect. D Biol. Crystallogr.* **2013**, *69*, 1223–1230. [CrossRef] [PubMed]
32. Van Genderen, E.; Clabbers, M.T.B.; Das, P.P.; Stewart, A.; Nederlof, I.; Barentsen, K.C.; Portillo, Q.; Pannu, N.S.; Nicolopoulos, S.; Gruene, T.; et al. Ab initio structure determination of nanocrystals of organic pharmaceutical compounds by electron diffraction at room temperature using a Timepix quantum area direct electron detector. *Acta Crystallogr. Sect. A* **2016**, *72*, 236–242.
33. Gemmi, M.; la Placa, M.G.I.; Galanis, A.S.; Rauch, E.F.; Nicolopoulos, S. Fast electron diffraction tomography. *J. Appl. Crystallogr.* **2015**, *48*, 718–727. [CrossRef]
34. Shi, D.; Nannenga, B.L.; Iadanza, M.G.; Gonen, T. Three-dimensional electron crystallography of protein microcrystals. *eLife* **2013**, *2*, e01345. [CrossRef]
35. Clabbers, M.T.B.; van Genderen, E.; Wan, W.; Wiegers, E.L.; Gruene, T.; Abrahams, J.P. Protein structure determination by electron diffraction using a single three-dimensional nanocrystal. *Acta Crystallogr. Sect. D* **2017**, *73*, 738–748. [CrossRef] [PubMed]
36. Clabbers, M.T.B.; Gruene, T.; Parkhurst, J.M.; Abrahams, J.P.; Waterman, D.G. Electron diffraction data processing with DIALS. *Acta Crystallogr. Sect. D Struct. Biol.* **2018**, *74*, 506–518. [CrossRef] [PubMed]
37. Giacovazzo, C.; Monaco, H.L.; Artioli, G.; Viterbo, D.; Milanesio, M.; Gilli, G.; Gilli, P.; Zanotti, G.; Ferraris, G.; Catti, M. *Fundamentals of Crystallography*; Oxford University Press: Oxford, UK, 2011; ISBN 9780199573653.

38. Dorset, D.L. Direct methods and refinement in electron and X-ray crystallography—Diketopiperazine revisited. *Z. Krist.* **2010**, *225*, 86–93. [CrossRef]
39. Palatinus, L. Structure Solution by Charge Flipping. In *Unitin Electron Crystallography and Powder Diffraction*; NATO Science for Peace and Security Series B: Physics and Biophysics; Kolb, U., Shankland, K., Meshi, L., David, W.I.F., Eds.; Springer: Dordrecht, The Netherlands, 2012; ISBN 9789400755796. [CrossRef]
40. Palatinus, L.; Van Der Lee, A. Symmetry determination following structure solution in P1. *J. Appl. Crystallogr.* **2008**, *41*, 975–984.
41. David, W.I.F.; Shankland, K.; van de Streek, J.; Pidcock, E.; Motherwell, W.D.S.; Cole, J.C. DASH: A program for crystal structure determination from powder diffraction data. *J. Appl. Crystallogr.* **2006**, *39*, 910–915. [CrossRef]
42. David, W.I.F.; Shankland, K.; Shankland, K.; Shankland, N. Routine determination of molecular crystal structures from powder diffraction data. *Chem. Commun.* **1998**, *8*, 931–932. [CrossRef]
43. Pagola, S.; Stephens, P.W.; Bohle, D.S.; Kosar, A.D.; Madsen, S.K. The structure of malaria pigment β-haematin. *Nature* **2000**, *404*, 307–310.
44. Favre-Nicolin, V.; Cerny, R. FOX, 'free objects for crystallography': A modular approach to ab initio structure determination from powder diffraction. *J. Appl. Crystallogr.* **2002**, *35*, 734–743. [CrossRef]
45. Černý, R.; Favre-Nicolin, V. Direct space methods of structure determination from powder diffraction: Principles, guidelines and perspectives. *Z. Krist.-Cryst. Mater.* **2007**, *222*, 105–113. [CrossRef]
46. Andreev, Y.G.; MacGlashan, G.S.; Bruce, P.G. Ab initio solution of a complex crystal structure from powder-diffraction data using simulated-annealing method and a high degree of molecular flexibility. *Phys. Rev. B* **1997**, *55*, 12011–12017. [CrossRef]
47. Harris, K.D.M.; Johnston, R.L.; Kariuki, B.M. The Genetic Algorithm: Foundations and Applications in Structure Solution from Powder Diffraction Data. *Acta Crystallogr. A* **1998**, *54*, 632–645. [CrossRef]
48. Shankland, K.; David, W.I.F.; Csoka, T. Crystal structure determination from powder diffraction data by the application of a genetic algorithm. *Z. Krist.-Cryst. Mater.* **1997**, *212*, 550–552. [CrossRef]
49. Deem, M.W.; Newsam, J.M. Framework crystal structure solution by simulated annealing: Test application to known zeolite structures. *J. Am. Chem. Soc.* **1992**, *114*, 7189–7198. [CrossRef]
50. Putz, H.; Schön, J.C.; Jansen, M. Combined method for ab initio structure solution from powder diffraction data. *J. Appl. Crystallogr.* **1999**, *32*, 864–870. [CrossRef]
51. Putz, H.; Brandenburg, K. Endeavour—Structure Solution from Powder Diffraction. Crystal Impact. Available online: https://www.crystalimpact.de/endeavour (accessed on 19 February 2021).
52. Hadermann, J.; Abakumov, A.M.; Tsirlin, A.A.; Filonenko, V.P.; Gonnissen, J.; Tan, H.; Verbeeck, J.; Gemmi, M.; Antipov, E.V.; Rosne, H. Direct space structure solution from precession electron diffraction data: Resolving heavy and light scatterers in $Pb_{13}Mn_9O_{25}$. *Ultramicroscopy* **2010**, *110*, 881–890. [CrossRef]
53. Gorelik, T.; Matveeva, G.; Kolb, U.; Schleuß, T.; Kilbinger, A.F.M.; van de Streek, J.; Bohle, A.; Brunklaus, G. H-bonding schemes of di- and tri-p-benzamides assessed by a combination of electron diffraction, X-ray powder diffraction and solid-state NMR. *CrystEngComm* **2010**, *12*, 1824–1832. [CrossRef]
54. Das, P.P.; Mugnaioli, E.; Nicolopoulos, S.; Tossi, C.; Gemmi, M.; Galanis, A.; Borodi, G.; Pop, M.M. Crystal Structures of Two Important Pharmaceuticals Solved by 3D Precession Electron Diffraction Tomography. *Org. Process Res. Dev.* **2018**, *22*, 1365–1372. [CrossRef]
55. Groom, C.R.; Bruno, I.J.; Lightfoot, M.P.; Ward, S.C. The Cambridge Structural Database. *Acta Crystallogr. Sect. B Struct. Sci. Cryst. Eng. Mater.* **2016**, *72*, 171–179. [CrossRef]
56. Faber, J. A New Relational Database Format for Powder Diffraction, Data Mining and Materials Characterization. *Suppl. J. Ceram. Soc. Jpn.* **2004**, S1434–S1438. [CrossRef]
57. Grazulis, S.; Chateigner, D.; Downs, R.T.; Yokochi, A.T.; Quiros, M.; Lutterotti, L.; Manakova, E.; Butkus, J.; Moeck, P.; Le Bail, A. Crystallography Open Database—An open-access collection of crystal structures. *J. Appl. Crystallogr.* **2009**, *42*, 726–729. [CrossRef]
58. Dill, K.A.; Phillips, A.T.; Rosen, J.B. Molecular Structure Prediction by Global Optimization. In *Developments in Global Optimization*; Bomze, I.M., Csendes, T., Horst, R., Pardalos, P.M., Eds.; Kluwer: Dordrecht, The Netherlands, 1997; pp. 217–234.
59. Calestani, G.; Rizzoli, C. Crystal structure of the $YBa_2Cu_3O_7$ superconductor by single-crystal X-ray diffraction. *Nature* **1987**, *328*, 606–607. [CrossRef]
60. Calamiotou, M.; Gantis, A.; Margiolaki, I.; Palles, D.; Siranidi, E.; Liarokapis, E. Phase separation, microstructure and superconductivity in the $Y_{1-x}Pr_xBa_2Cu_3O_y$ compounds. *J. Phys. Condens. Matter* **2008**, *20*, 395224. [CrossRef]
61. Paulus, E.F.; Miehe, G.; Fuess, H.; Yehia, I.; Löchner, U. The crystal structure of $BaCuO_2$. *J. Solid State Chem.* **1991**, *90*, 17–26. [CrossRef]
62. Kuntzinger, S.; Ghermani, N.E. Electron density distribution and Madelung potential in α-spodumene, $LiAl(SiO_3)_2$, from two-wavelength high-resolution X-ray diffraction data. *Acta Crystallogr. Sect. B* **1999**, *55*, 273–284. [CrossRef]
63. Redhammer, G.J.; Roth, G. Structural variation and crystal chemistry of $LiMe^{3+}Si_2O_6$ clinopyroxenes Me^{3+} = Al, Ga, Cr, V, Fe, Sc and In. *Z. Krist.-Cryst. Mater.* **2004**, *219*, 278–294. [CrossRef]
64. Redhammer, G.J.; Roth, G. Structural changes upon the temperature dependent C2/c → P21/c phase transition in $LiMe^{3+}Si_2O_6$ clinopyroxenes, Me = Cr, Ga, Fe, V, Sc and In. *Z. Krist.-Cryst. Mater.* **2004**, *219*, 585–605. [CrossRef]
65. Iezzi, G.; Bromiley, G.D.; Cavallo, A.; Das, P.P.; Karavassili, F.; Margiolaki, I.; Stewart, A.A.; Tribaudino, M.; Wright, J.P. Solid solution along the synthetic $LiAlSi_2O_6$-$LiFeSi_2O_6$ (spodumene-ferri-spodumene) join: A general picture of solid solutions, bond

lengths, lattice strains, steric effects, symmetries, and chemical compositions of Li clinopyroxenes. *Am. Mineral.* **2016**, *101*, 2498–2513. [CrossRef]
66. Bertrand, C.; Galez, P.; Jorda, J.L.; Gladyshevskii, R.E. The Pr $(Ba_{1-x}Pr_x)_2$ Cu_3O_{7+d} solid solution. A crystal structure and phase diagram study. *Phys. C Supercond.* **1999**, *321*, 151–161. [CrossRef]
67. Kipka, R.; Müller-Buschbaum, H. Über Oxocuprate, XX Ein Erdalkalioxocuprat(II) mit geschlossenen Baugruppen: $BaCuO_2$/about Oxocuprates, XX Alkaline-earth Oxocuprate(II) with Closed Structural Groups: $BaCuO_2$. *Z. Nat. B* **1977**, *32*, 121–123. [CrossRef]
68. Plana-Ruiz, S. Development & Implementation of an Electron Diffraction Approach for Crystal Structure Analysis—TUprints (TU-Darmstadt.De). Available online: https://www.tesisenred.net/handle/10803/670887#page=1 (accessed on 12 January 2022).
69. Plana-Ruiz, S.; Krysiak, Y.; Portillo, J.; Alig, E.; Estradé, S.; Peiró, F.; Kolb, U. Fast-ADT: A fast and automated electron diffraction tomography setup for structure determination and refinement. *Ultramicroscopy* **2020**, *211*, 112951. [CrossRef]
70. Plana-Ruiz, S.; Portillo, J.; Estradé, S.; Peiró, F.; Kolb, U.; Nicolopoulos, S. Quasi-parallel precession diffraction: Alignment method for scanning transmission electron microscopes. *Ultramicroscopy* **2018**, *193*, 39–51. [CrossRef] [PubMed]
71. Smeets, S.; Wang, B.; Cichocka, M.O.; Ångström, J.; Wan, W. Continuous rotation electron diffraction data for zeolite SSZ-27. *Instamatic Zenodo* **2018**. [CrossRef]
72. Gorelik, T.E.; Stewart, A.A.; Kolb, U. Structure solution with automated electron diffraction tomography data: Different instrumental approaches. *J. Microsc.* **2011**, *244*, 325–331. [CrossRef] [PubMed]
73. Kolb, U.; Krysiak, Y.; Plana-Ruiz, S. Automated electron diffraction tomography—Development and applications. *Acta Crystallogr. Sect. B Struct. Sci. Cryst. Eng. Mater.* **2019**, *75*, 463–474. [CrossRef]
74. Cascarano, G.L.; Giacovazzo, C.; Carrozzini, B. Crystal structure solution via precession electron diffraction data: The BEA algorithm. *Ultramicroscopy* **2010**, *111*, 56–61. [CrossRef]
75. Burla, M.C.; Caliandro, R.; Carrozzini, B.; Cascarano, G.L.; Cuocci, C.; Giacovazzo, C.; Mallamo, M.; Mazzone, A.; Polidori, G. Crystal structure determination and refinement via SIR2014. *J. Appl. Crystallogr.* **2015**, *48*, 306–309. [CrossRef]
76. Petříček, V.; Dušek, M.; Palatinus, L. Crystallographic Computing System JANA2006: General features. *Z. Krist.-Cryst. Mater.* **2014**, *229*, 345–352. [CrossRef]
77. Harris, K.D.M.; Tremayne, M. Crystal Structure Determination from Powder Diffraction Data. *Chem. Mater.* **1996**, *8*, 2554–2570. [CrossRef]
78. Louër, D. Advances in Powder Diffraction Analysis. *Acta Cryst. Sect. A* **1998**, *54*, 922–933. [CrossRef]
79. Kaplow, R.; Rowe, T.A.; Averbach, B.L. Atomic Arrangement in Vitreous Selenium. *Phys. Rev.* **1968**, *168*, 1068–1079. [CrossRef]
80. McGreevy, R.L. 6—Reverse Monte Carlo Methods for Structural Modelling. In *Computer Modeling in Inorganic Crystallography*; Catlow, C.R.A., Ed.; Academic Press: London, UK, 1997; pp. 151–184. [CrossRef]
81. Pareto, V. *Cours d'Économie Politique*; F. Rouge: Laussane, Switzerland, 1896.
82. De Leeuw, S.W.; Perram, J.W.; Smith, E.R. Simulation of electrostatic systems in periodic boundary conditions. I. Lattice sums and dielectric constants. *Proc. R. Soc. A* **1980**, *373*, 27–57. [CrossRef]
83. Kirkpatrick, S.; Gelatt, C.D.; Vecchi, M.P. Optimization by Simulated Annealing. *Science* **1983**, *220*, 671–680. [CrossRef] [PubMed]
84. Hannemann, A.; Hundt, R.; Schoen, J.C.; Jansen, M. A New Algorithm for Space-Group Determination. *J. Appl. Cryst.* **1998**, *31*, 922–928. [CrossRef]
85. Hundt, R.; Schoen, J.C.; Hannemann, A.; Jansen, M. Determination of Symmetries and Idealized Cell Parameters for Simulated Structures. *J. Appl. Cryst.* **1999**, *32*, 413–416. [CrossRef]
86. Prince, E. (Ed.) *International Tables for Crystallography. Vol. C: Mathematical, Physical and Chemical Tables*, 3rd ed.; Kluwer Academic Publishers: Dordrecht, The Netherlands, 2004.
87. Moeck, P.; (Portland State University, Portland, OR, USA). Personal communication, 2008.
88. Ge, M.; Yang, T.; Wang, Y.; Carraro, F.; Liang, W.; Doonan, C.; Falcaro, P.; Zheng, H.; Zou, X.; Huang, Z. On the completeness of three-dimensional electron diffraction data for structural analysis of metal-organic frameworks. *Faraday Discuss.* **2021**, *231*, 66–80. [CrossRef]
89. Wennmacher, J.T.C.; Zaubitzer, C.; Li, T.; Bahk, Y.K.; Wang, J.; van Bokhoven, J.A.; Gruene, T. 3D-structured supports create complete data sets for electron crystallography. *Nat. Commun.* **2019**, *10*, 3316. [CrossRef]
90. Palatinus, L.; Petříček, V.; Corrêa, C.A. Structure refinement using precession electron diffraction tomography and dynamical diffraction: Theory and implementation. *Acta Crystallogr. Sect. A* **2015**, *71*, 235–244. [CrossRef]
91. Momma, K.; Izumi, F. VESTA 3 for three-dimensional visualization of crystal, volumetric and morphology data. *J. Appl. Crystallogr.* **2011**, *44*, 1272–1276. [CrossRef]
92. Brown, I.D. 14—The Bond-Valence Method: An Empirical Approach to Chemical Structure and Bonding. *Ind. Chem. Libr.* **1981**, *2*, 1–30. [CrossRef]
93. Brown, I.D.; Altermatt, D. Bond-valence parameters obtained from a systematic analysis of the Inorganic Crystal Structure Database. *Acta Crystallogr. Sect. B* **1985**, *41*, 244–247. [CrossRef]

MDPI AG
Grosspeteranlage 5
4052 Basel
Switzerland
Tel.: +41 61 683 77 34

Symmetry Editorial Office
E-mail: symmetry@mdpi.com
www.mdpi.com/journal/symmetry

www.ingramcontent.com/pod-product-compliance
Lightning Source LLC
Chambersburg PA
CBHW062043220926
51850CB00010B/2644